The Uncertainty in Physical Measurements

Paolo Fornasini

The Uncertainty in Physical Measurements

An Introduction to Data Analysis
in the Physics Laboratory

Paolo Fornasini
Department of Physics
University of Trento
Italy
paolo.fornasini@unitn.it

ISBN: 978-0-387-78649-0   e-ISBN: 978-0-387-78650-6
DOI: 10.1007/978-0-387-78650-6

Library of Congress Control Number: 2008931297

© 2008 Springer Science+Business Media, LLC
All rights reserved. This work may not be translated or copied in whole or in part without the written permission of the publisher (Springer Science+Business Media, LLC, 233 Spring Street, New York, NY 10013, USA), except for brief excerpts in connection with reviews or scholarly analysis. Use in connection with any form of information storage and retrieval, electronic adaptation, computer software, or by similar or dissimilar methodology now known or hereafter developed is forbidden.
The use in this publication of trade names, trademarks, service marks, and similar terms, even if they are not identified as such, is not to be taken as an expression of opinion as to whether or not they are subject to proprietary rights.

Printed on acid-free paper

9 8 7 6 5 4 3 2 1

springer.com

To my wife Chiara

Preface

The scientific method is based on the measurement of different physical quantities and the search for relations between their values. All measured values of physical quantities are, however, affected by uncertainty. Understanding the origin of uncertainty, evaluating its extent, and suitably taking it into account in data analysis, are fundamental steps for assessing the global accuracy of physical laws and the degree of reliability of their technological applications.

The introduction to uncertainty evaluation and data analysis procedures is generally made in laboratory courses for freshmen. During my long-lasting teaching experience, I had the feeling of some sort of gap between the available tutorial textbooks, and the specialized monographs. The present work aims at filling this gap, and has been tested and modified through a feedback interaction with my students for several years. I have tried to maintain as much as possible a tutorial approach, that, starting from a phenomenological introduction, progressively leads to an accurate definition of uncertainty and to some of the most common procedures of data analysis, facilitating the access to advanced monographs. This book is mainly addressed to undergraduate students, but can be a useful reference for researchers and for secondary school teachers.

The book is divided into three parts and a series of appendices.

Part I is devoted to a phenomenological introduction to measurement and uncertainty. In Chap. 1, the direct and indirect procedures for measuring physical quantities are distinguished, and the unavoidability of uncertainty in measurements is established from the beginning. Measuring physical quantities requires the choice of suitable standard units, and Chap. 2 is dedicated to the International System of units and to dimensional analysis. To perform measurements, suitable instruments are necessary; the basic properties of instruments are presented in Chap. 3, including the characteristics of static and dynamic performance. Chap. 4 plays a central role; here, the different possible causes of uncertainty are thoroughly explored and compared, and the methodologies for quantitatively evaluating and expressing the uncertainty are explained. The phenomenological introduction of the normal and uniform distributions naturally leads to the demand for a more formal probabilistic approach.

To such an approach, Part II is dedicated. In Chap. 5, the basic concepts of probability theory are presented: sample space, events, definitions of probability, sum and product of events. The theory of probability is further developed in Chap. 6, through the formalism of random variables; the general properties of the distributions of random variables are introduced, and attention is focused on the distributions most frequently encountered in physics: binomial, Poisson, normal, uniform, and Cauchy–Lorentz. Chap. 7 is devoted to some basic statistical concepts and tools: parent and sample populations, estimate of population parameters, and the maximum likelihood criterion.

In Part III, some common data analysis procedures are introduced. Chap. 8 is dedicated to the propagation of uncertainty in indirect measurements. Chap. 9 introduces the distinction between probability and confidence, and presents some relevant applications of the confidence level and the Student distribution. In Chap. 10, the correlation between physical quantities is quantitatively studied by introducing the linear correlation coefficient and the procedures of regression based on the least squares method. Finally, an introduction to the chi square statistical test is made in Chap. 11.

Part IV contains a set of appendices. A clever presentation of data increases the effectiveness of analysis procedures, and guarantees accuracy in communicating the results to other researchers. Appendix A is dedicated to the treatment of significant digits and the use of tables, graphs, and histograms. Appendix B is dedicated to the International System of Units (SI) and to other frequently used systems of units. Appendix C contains some useful tables: the Greek alphabet, a list of selected constants of physics, and the integrals of the probability distributions introduced in previous chapters. Mathematical technicalities have been avoided as much as possible in the main text of the book. Some useful demonstrations can, however, be found in Appendix D by interested readers. The comprehension of theoretical concepts is greatly facilitated by the possibility of practical applications. Several problems are proposed at the end of some chapters. Solving statistical problems is, however, much more effective if they refer to real experiments. Appendix E contains the description of some simple experiments, particularly suited to illustrate the data analysis procedures introduced in this book. The experiments are based on cheap and easily available instrumentation, and their effectiveness has been tested by many classes of students.

I am indebted to a large number of colleagues and students for stimulating discussions. Let me here remember in particular M. Grott, G. Prodi, and L. Tubaro, for their invaluable advice.

Povo

Paolo Fornasini

January 2008

Contents

Preface VII

Part I Measurements and Uncertainty

1 Physical Quantities 3
  1.1 Methods of Observation and Measurement 3
  1.2 Physical Quantities 5
  1.3 Direct and Indirect Measurement 6
  1.4 Time Dependence of Physical Quantities 8
  1.5 Counting of Random Events 10
  1.6 Operative Definition of Physical Quantities 11
  1.7 The Experimental Method 12

2 Measurement Units 13
  2.1 Base and Derived Quantities 13
  2.2 Measurement Standards 14
  2.3 The International System of Units (SI) 15
  2.4 Other Systems of Units 18
  2.5 Dimensional Analysis 20
  Problems 24

3 Measuring Instruments 27
  3.1 Functional Elements 27
  3.2 Classifications of Instruments 29
  3.3 Static Characteristics of Instruments 31
  3.4 Accuracy of an Instrument 35
  3.5 Dynamical Behavior of Instruments 37
  3.6 Counters 43

4 Uncertainty in Direct Measurements 45
  4.1 Causes of Uncertainty 45
  4.2 Measurement Resolution 46
  4.3 Random Fluctuations 48
  4.4 Systematic Errors 61

	4.5	Summary and Comparisons	68
		Problems	74

Part II Probability and Statistics

5	**Basic Probability Concepts**		79
	5.1	Random Phenomena	79
	5.2	Sample Space. Events	81
	5.3	Probability of an Event	82
	5.4	Addition and Multiplication of Events	87
	5.5	Probability of the Sum of Events	89
	5.6	Probability of the Product of Events	91
	5.7	Combinatorial Calculus	94
		Problems	96
6	**Distributions of Random Variables**		99
	6.1	Binomial Distribution	99
	6.2	Random Variables and Distribution Laws	104
	6.3	Numerical Characteristics of Distributions	108
	6.4	Poisson Distribution	115
	6.5	Normal Distribution	121
	6.6	Meaning of the Normal Distribution	126
	6.7	The Cauchy–Lorentz Distribution	130
	6.8	Multivariate Distributions	132
		Problems	136
7	**Statistical Tools**		139
	7.1	Parent and Sample Populations	139
	7.2	Sample Means and Sample Variances	143
	7.3	Estimation of Parameters	147
		Problems	152

Part III Data Analysis

8	**Uncertainty in Indirect Measurements**		155
	8.1	Introduction to the Problem	155
	8.2	Independent Quantities, Linear Functions	156
	8.3	Independent Quantities, Nonlinear Functions	159
	8.4	Nonindependent Quantities	163
	8.5	Summary	167
		Problems	168

9 Confidence Levels 169
- 9.1 Probability and Confidence 169
- 9.2 The Student Distribution 173
- 9.3 Applications of the Confidence Level 174
- Problems 176

10 Correlation of Physical Quantities 177
- 10.1 Relations Between Physical Quantities 177
- 10.2 Linear Correlation Coefficient 179
- 10.3 Linear Relations Between Two Quantities 181
- 10.4 The Least Squares Method 186
- Problems 192

11 The Chi Square Test 193
- 11.1 Meaning of the Chi Square Test 193
- 11.2 Definition of Chi Square 194
- 11.3 The Chi Square Distribution 198
- 11.4 Interpretation of the Chi Square 202
- Problems 203

Part IV Appendices

A Presentation of Experimental Data 207
- A.1 Significant Digits and Rounding 207
- A.2 Tables 210
- A.3 Graphs 212
- A.4 Histograms 216

B Systems of Units 219
- B.1 The International System of Units (SI) 219
- B.2 Units Not Accepted by the SI 223
- B.3 British Units 224
- B.4 Non-SI Units Currently Used in Physics 225
- B.5 Gauss cgs Units 226

C Tables 227
- C.1 Greek Alphabet 227
- C.2 Some Fundamental Constants of Physics 227
- C.3 Integrals of the Standard Normal Distribution 229
- C.4 Integrals of the Student Distribution 233
- C.5 Integrals of the Chi Square Distribution 235
- C.6 Integrals of the Linear Correlation Coefficient Distribution ... 237

D Mathematical Complements ... 239
- D.1 Response of Instruments: Differential Equations ... 239
- D.2 Transformed Functions of Distributions ... 243
- D.3 Moments of the Binomial Distribution ... 245
- D.4 Moments of the Uniform Distribution ... 246
- D.5 Moments of the Poisson Distribution ... 247
- D.6 Moments of the Normal Distribution ... 248
- D.7 Parameters of the Cauchy Distribution ... 252
- D.8 Theorems on Means and Variances ... 253

E Experiments ... 255
- E.1 Caliper and Micrometer ... 255
- E.2 Simple Pendulum: Measurement of Period ... 258
- E.3 Helicoidal Spring: Elastic Constant ... 261
- E.4 Helicoidal Spring: Oscillations ... 266
- E.5 Simple Pendulum: Dependence of Period on Length ... 269
- E.6 Simple Pendulum: Influence of Mass and Amplitude ... 274
- E.7 Time Response of a Thermometer ... 277

Suggested Reading ... 283

Index ... 285

Part I

Measurements and Uncertainty

1 Physical Quantities

The great power of the scientific method relies on the possibility of singling out several measurable properties, the physical quantities, and of finding stable relations between the measured values of different physical quantities.

What do we mean by physical quantity? Why, for example, are length and mass physical quantities, and taste and smell are not? The very definition of physical quantities, as well as their practical use, is strictly connected to the definition of a measurement procedure, which allows us to establish a correspondence between physical quantities and numbers.

Every practical measurement entails a degree of uncertainty in its result. Otherwise stated, uncertainty is an integral part of every measure. The ability of evaluating the measurement uncertainty is fundamental both in scientific research, to establish the validity limits of theories, and in technological applications, to assess the reliability of products and procedures.

1.1 Methods of Observation and Measurement

To clarify the concept of physical quantity, it is useful to distinguish the different methods that are used to study natural phenomena, and classify them in order of increasing complexity and power.

Morphological Method

The simplest method consists of the sensorial detection of some properties of objects or phenomena, and possibly their registration and description, by means of drawings, photographs, and so on.

Example 1.1. The study of anatomy is generally done by means of photographs, drawings, and movies.

Example 1.2. Many chemical substances can be identified, by a skilled person, through their color, brilliance, smell, taste, and so on.

Classificatory Method

Progress is made when there is the possibility of partitioning a set of objects or phenomena into classes, according to the fulfillment of well-defined requisites. Two objects or phenomena belong to the same class if, and only if, they share at least one property.

Example 1.3. Zoology and botanics are based on complex and articulated classifications. For example, vertebrate animals are divided into five classes: mammals, birds, reptiles, amphibians, and fish. Every class is in turn divided into many orders, families, and species, according to a descending hierarchy.

Example 1.4. The relation of congruence of segments is an equivalence relation, and allows one to group the segments of space into classes, called *lengths*. Note, however, that at this level length cannot be considered a physical quantity as yet.

Comparative Method

Further progress is represented by the possibility of introducing an order relation, say a criterion for deciding whether, of two objects or phenomena, the first one possesses a given property in a smaller, equal, or larger degree than the second one. If the order relation is transitive, one can establish a correspondence between the degrees of the physical property and a set of numbers, such that the order relation is preserved. A property for which a comparative method can be defined is a *physical quantity*.

Example 1.5. The Mohs scale for the hardness of minerals is based on the following criterion: mineral A is harder than mineral B, if A can scratch B. This criterion establishes a transitive order relation: Hardness is a physical quantity. The Mohs scale lists ten minerals in order of growing hardness, associated with ten numbers: 1–talc, 2–gypsum, 3–calcite, 4–fluorite, 5–apatite, 6–orthoclase, 7–quartz, 8–topaz, 9–corundum, and 10–diamond. The choice of numbers is arbitrary, provided they are consistent with the order relation.

Example 1.6. By superposing two segments, one can decide if one of them is shorter or longer than the other, thus operatively introducing a transitive order relation. With every length, one can associate a number, arbitrarily chosen, provided the order relation is preserved. One can now say that length is a physical quantity.

Example 1.7. The thermal states of two objects A and B can be compared by bringing the two objects successively in contact with the same thermoscope. (A mercury thermoscope consists of a glass bulb filled with mercury, connected to a thin glass tube; it is like a common thermometer, without scale.) The comparison allows one to establish a transitive order relation. We can say that A has higher temperature than B if A induces a stronger dilatation of the thermoscope mercury. With the thermal state one can associate a number, arbitrarily chosen, provided the order relation is preserved. Temperature is a physical quantity.

Quantitative Method

In some cases, in addition to a transitive order relation, it is also possible to define a composition rule having the same properties of the addition of numbers. One can then establish a correspondence between the degrees of the physical property and the set of real numbers, reproducing not only the order relation but also the additive structure.

Example 1.8. An additive composition rule for segments can be introduced as follows. Let $a = AA'$ and $b = BB'$ be two segments. The composition consists of aligning the two segments on the same straight line, in such a way that the extremum A' of segment a coincides with the extremum B of segment b. The segment sum is $a + b = AB'$.

The quantitative method is the basis of most scientific and technical measurements. It allows the description of natural phenomena through mathematical formalisms. Notice, however, that an additive composition rule cannot be defined for all physical quantities (typical examples are hardness and temperature).

Statistical Methods

When studying very large populations of objects or events, it is often possible to describe some of their average properties by a limited number of parameters, utilizing statistical methods. For example, statistical methods are used in physics to give an interpretation of thermodynamic quantities (such as pressure, internal energy, temperature, and so on) in terms of the average behavior of a very large number of atoms or molecules. As shown in the following chapters, statistical methods play a fundamental role in the treatment of uncertainties of physical quantities.

1.2 Physical Quantities

According to Sect. 1.1, a *physical quantity* is a property of an object or phenomenon for which one can define a transitive order relation. For many physical quantities, an additive rule of composition can be defined as well. Physical quantities can thus be grouped into two sets.

(a) Additive quantities, for which one can define both a transitive order relation and an additive composition rule. Length, time interval, mass, speed, and force are examples of additive quantities.

(b) *Nonadditive quantities*, for which one can define a transitive order relation but not an additive composition rule. Temperature and hardness are examples of nonadditive quantities.

Measurement is the experimental procedure by which a number, the *measure*, is associated with every value of a physical quantity. Different measurement methodologies have to be developed, according to whether one deals with an additive or a nonadditive quantity.

1.3 Direct and Indirect Measurement

Two different measurement methods have to be distinguished, direct and indirect measurement.

Direct Measurement

Let us consider an additive quantity $\mathcal{G}$; for the sake of concreteness, let it be the length of a rod. The direct measurement of the quantity $\mathcal{G}$ can be decomposed into the following sequence of logical steps.

(a) Construction or choice of a unit standard $\mathcal{U}$
(b) Composition of unit standards, $\sum \mathcal{U}_i$
(c) Check of the correspondence between $\mathcal{G}$ and a sum $n\mathcal{U}$ of unit standards
(d) Counting of the n unit standards

According to this logical scheme, the *measure* $X(\mathcal{G})$ of the quantity $\mathcal{G}$ is the ratio between the quantity $\mathcal{G}$ and the unit standard $\mathcal{U}$:

$$X(\mathcal{G}) = \mathcal{G}/\mathcal{U}. \tag{1.1}$$

The result of a direct measurement is thus written as $\mathcal{G} = X\,\mathcal{U}$, the number X being the measure and $\mathcal{U}$ being the unit. Examples: for a length, $d = 5$ meters; for a time interval, $\Delta t = 7$ seconds; for a mass, $m = 2$ kilograms.

The direct measurement is the operative realization of the quantitative method, and is possible only for additive quantities.

Example 1.9. Measurement of the length of an object by a ruler. The unit standard is the distance between two adjacent cuts (1 mm); the composition of unit standards has been made when the cuts have been engraved on the rule; the check of correspondence is performed by bringing to coincidence the sides of the object with the cuts on the ruler; the count of standards is facilitated by the numbers engraved on the ruler.

Example 1.10. Measurement of a mass by an equal-arm balance. The standard is a unit mass; the composition of standards is made by placing several unit masses on the same scale-pan of the balance; the check of correspondence consists of calibrating the number of unit masses in order to obtain equilibrium between the two scale-pans.

Uncertainty in Direct Measurements

Let us now analyze in more detail the meaning of (1.1), starting from purely mathematical arguments, and considering then the experimental factors that can influence the measurement.

Only seldom will the quantity $\mathcal{G}$ correspond to an integer multiple $n\mathcal{U}$ of the unit standard. In general, the measure $X = \mathcal{G}/\mathcal{U}$ is not an integer number n. If the unit $\mathcal{U}$ is supposed indefinitely divisible into submultiples, one could guess that the measure $X = \mathcal{G}/\mathcal{U}$ is a rational number m/n. It is, however, well known that incommensurate quantities exist (such as the side and the diagonal of a square), whose ratio is an irrational number. As a consequence, the measure of a physical quantity is in principle a real number r:

$$X(\mathcal{G}) = \mathcal{G}/\mathcal{U} = r. \tag{1.2}$$

In an actual experimental measurement, one always deals with instruments whose unit $\mathcal{U}$ cannot be made arbitrarily small. As a consequence, the check of identity between the quantity $\mathcal{G}$ and a sum of unit standards only allows one to state that

$$n\mathcal{U} < \mathcal{G} < (n+1)\mathcal{U}, \tag{1.3}$$

say to define an interval of values of width $\mathcal{U}$, which includes $\mathcal{G}$. Otherwise stated, the result of a direct measurement is always represented by a finite interval of possible values. The width of this interval represents an *uncertainty* of the measure.

In principle, one could think that the uncertainty can be reduced below any predetermined value by suitably reducing the unit $\mathcal{U}$. In practice, the reduction of $\mathcal{U}$ is limited by technical difficulties. Anyway, as shown in subsequent chapters, other causes, depending on both random fluctuations and systematic errors, contribute to the uncertainty and become predominant when $\mathcal{U}$ is sufficiently small.

It is a result of long-term experience that uncertainty in measurements can never be completely eliminated. Uncertainty is then an integral part of every measure, and always has to be carefully evaluated. The measure of any physical quantity must always contain information about its uncertainty. The standard expression of a physical measure is

$$\mathcal{G} = (X_0 \pm \delta X)\mathcal{U}, \tag{1.4}$$

where X_0 is the central value of the measure, and δX is the uncertainty, here taken as the half-width of the uncertainty interval.

Chapter 4 is devoted to the evaluation of uncertainty in various realistic situations, and to the definition of its standard expression.

Indirect Measurement

Indirect measurement is the procedure by which the measure $X(\mathcal{G})$ of a quantity $\mathcal{G}$ is obtained through analytical relations from the measures $Y(\mathcal{A}), Z(\mathcal{B}), \ldots$ of other quantities $\mathcal{A}, \mathcal{B}, \ldots$ directly measured.

Quantities that could in principle be directly measured are often indirectly measured, for convenience or because of the practical difficulty of direct measurements.

Example 1.11. Velocity is in principle directly measurable (if relativistic effects can be neglected). In general, however, it is indirectly measured, for example, as a ratio between a length and a time interval.

Example 1.12. Astronomic distances cannot be directly measured for practical reasons. Their indirect measurement is based on direct measurements of angles and of a small distance.

Indirect measurement is necessary for nonadditive quantities, such as temperature. When using calibrated instruments, however, the procedure of indirect measurement is not evident to the user.

Example 1.13. Temperature cannot be directly measured, because an additive composition law cannot be defined. Different thermometric scales connect temperature to a directly measurable quantity, such as length, pressure, electric resistance, and so on. In a mercury thermometer, the measurement of a temperature variation ΔT is connected to the direct measurement of the expansion $\Delta \ell$ of mercury; the reading is, however, directly given in temperature units.

The uncertainty, which characterizes every direct measure, obviously propagates to the indirect measures. The propagation of uncertainty from direct to indirect measures is considered in Chap. 8.

1.4 Time Dependence of Physical Quantities

Among physical quantities, time plays a particular role. Frequently, one is interested in the time variations of a given quantity $\mathcal{G}$, say in the behavior of the function $X(t)$, where X is the measure of $\mathcal{G}$ and t is time. The most important kinds of time dependence are described below, and schematically depicted in Fig. 1.1.

A quantity $\mathcal{G}$ is *constant* when its value $X(\mathcal{G})$ does not change with time.

Example 1.14. The force of gravitational interaction between two masses m and M at a distance r is $F = GmM/r^2$. The value of the gravitational constant G does not depend on the site and time of measurement. G is a fundamental constant of physics.

1.4 Time Dependence of Physical Quantities

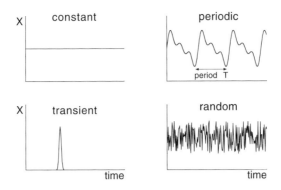

Fig. 1.1. Time dependence of physical quantities: constant, periodic, transient, and random quantities.

Example 1.15. The period T of oscillation of a pendulum depends on the amplitude of oscillation, which decreases in time as an effect of friction. However, if the oscillations are small enough, the dependence of period on amplitude is weak and the friction damping is slow. In this situation, the period can be considered constant with good approximation.

A quantity X is *periodic* with period T if, for each time value t,

$$X(t+T) = X(t) \,. \tag{1.5}$$

Example 1.16. Within the approximation of small oscillations, the angle θ of an oscillating pendulum varies periodically with time. The dependence of θ on time is given by a sinusoidal law: $\theta = \theta_0 \sin(2\pi t/T + \phi_0)$, where θ_0 is the amplitude of oscillation, T the period, and ϕ_0 the phase for $t=0$.

The periodic character of some natural phenomena – such as rotation or revolution of planets, oscillation of pendulums, and vibrations of quartz crystals – allows one to establish the procedures for measuring the time intervals, thus making possible the very definition of *time interval*.

A physical quantity is *impulsive* or *transient* when its value is different from zero only in a finite time interval.

Example 1.17. When two solid bodies collide, the force of interaction varies very quickly in time within the short time interval during which the two bodies are in contact. The force has an impulsive character.

A physical quantity is *random* or *casual* when its value continuously varies in time, in a nonperiodic way.

Example 1.18. The molecules of a gas continuously move and collide with the container walls and with other molecules. At each collision, the velocity of a molecule changes direction and intensity in an unpredictable way. The velocity of a molecule is thus a random quantity.

The methodology of measurement of a physical quantity, as well as the choice of the most suitable instruments, depends on its time dependence. To measure a constant quantity, the available time is virtually unlimited; one can obtain reliable measures with relatively simple instruments, by exploiting the possibility of repeating the measurement many times and of modifying the procedures. The measurement of time-dependent quantities requires instead fast enough instruments and procedures (see Chap. 3).

1.5 Counting of Random Events

The direct measurement of a physical quantity requires counting a number n of identical standard units $\mathcal{U}$. A simple example is given by the measurement of a time interval $\Delta t = t_2 - t_1$ by a clock with 1 second resolution, $\mathcal{U} = 1$ s. The measurement consists of counting the number of seconds from the initial time t_1 to the final time t_2. The signals produced by the clock at every second are events regularly spaced in time (Fig. 1.2, left).

Fig. 1.2. Regularly spaced events (left) and random events (right).

A completely different situation is encountered when dealing with *random events*, say events dispersed in a disordered and unpredictable way in time or space (Fig. 1.2, right). A typical example of random events is represented by cosmic rays, say high-energy particles that continuously arrive on the Earth's surface. Cosmic rays can be detected and counted by means of suitable instruments, such as Geiger counters (Sect. 3.6). The arrival of a particle at the entrance window of a Geiger counter is a random event, and the time of arrival is completely unpredictable. The sequence of detection of cosmic rays is disordered and causal, as in Fig. 1.2, right.

Many other phenomena of physics give rise to random events, such as the decay of radioactive isotopes, the emission of photons (light quanta) by atoms, and elementary particle collisions.

One could think that randomness prevents us from performing quantitative studies. However, if a sufficiently high number of random events is considered, it is possible to extract some stable *average properties* from the irregular time or space sequences. These average properties represent physical quantities.

Example 1.19. Let us consider a sample of radioactive isotopes. A radioactive decay is a random event, and the time sequence of decays is random.

However, if a large enough number of decays is observed, it is possible to note a regularity in their average behavior. If N_0 isotopes are present at time $t = 0$, their average number reduces in time according to the exponential law: $N(t) = N_0 \exp(-\alpha t)$, where α is the disintegration constant. The disintegration constant is a physical quantity, whose value characterizes the behavior of a specific isotope.

Counting of random events requires suitable instruments and suitable statistical techniques of data analysis, which are considered in the next chapters, and is intrinsically affected by uncertainty.

1.6 Operative Definition of Physical Quantities

As shown in previous sections, a physical quantity is by definition a property of objects or phenomena for which one can define a transitive order relation, and possibly also an additive composition rule. The existence of an order relation, and possibly of a composition rule, is in turn necessary for measuring physical quantities. For example, the procedure of direct measurement requires that a practical rule for summing up the standard units $\mathcal{U}$ is given. The very concept of physical quantity is thus intrinsically connected to the measurement procedures.

This standpoint naturally leads to the *operative definition* of physical quantities. Physical quantities (length, mass, time, force, and so on) are not defined in terms of abstract properties, but in terms of realistic procedures: the definition of each physical quantity consists of the detailed description of the procedures for its measurement.

The development of science and technology has led to a progressive extension of the use of physical quantities outside the field of everyday common experience. As a consequence, the values of physical quantities can span over many orders of magnitude. For example, let us consider two typical lengths, one of atomic physics and one of astronomy: the radius of the first electronic orbit of the atomic Bohr model is about 5.3×10^{-11} m, whereas the average radius of the orbit of Pluto is $5.9 \times 10^{+12}$ m.

It is evident that neither the lengths at atomic scale nor the lengths of astronomical interest can be measured through a direct comparison with a unit standard. In general, the same physical quantity can require different measurement procedures for different orders of magnitude of its values. Different measurement procedures in turn correspond to different operative definitions. To guarantee the possibility of referring to the same physical quantity in spite of the difference in measurement procedures, it is necessary that the different operative definitions are consistent: if two measurement procedures can be adopted within the same interval of values, they must give the same results. A physical quantity is thus defined by the class of all its possible and consistent operative definitions.

1.7 The Experimental Method

Physics is not limited to the observation of natural phenomena and to the measurement of physical quantities. The great power of the scientific method relies on the ability of finding correlations between different physical quantities, so as to establish laws and build up theories of general validity.

The search for correlations between physical quantities is not a simple task. Natural phenomena generally appear in complex forms, characterized by the simultaneous presence and mutual influence of many factors: a typical example is the ubiquitous presence of friction that prevented us, for many centuries, from understanding the relation of proportionality existing between force and acceleration, which is now at the basis of classical mechanics. Many important phenomena even escape a direct sensorial perception and can be detected only by means of suitable instrumentation (let us here mention only electromagnetic phenomena).

The search for correlations between physical quantities is founded on the *experimental method*, introduced by Galileo Galilei (1564–1642). A scientist is not confined to a passive observation of natural phenomena; he or she reproduces instead the phenomena in a controlled way in the laboratory, systematically modifying the factors that can affect their development. By this procedure, one can reduce or even eliminate secondary factors, and isolate the fundamental aspects of phenomena. One can thus detect simple empirical relations between physical quantities. For example, it was the progressive reduction of friction on a body moving along an inclined plane that led Galileo to hypothesize the relation of proportionality between force and acceleration.

The relations, experimentally established, between physical quantities are the basis for the development of scientific theories. Scientific theories, in turn, allow one to make previsions on the evolution of more complex phenomena. The validity of every scientific theory is corroborated by the experimental verification of its expectations.

The experimental method relies on a careful choice of the physical quantities used to describe natural phenomena, and on their rigorous operative definition. It also requires critical sense, technical ability, and some fantasy, in order to find simple and reproducible correlations between physical quantities.

A good researcher should be able to correctly evaluate the reliability of his results. Actually, the measurement of whichever physical quantity is affected by uncertainty, due to the instruments and procedures of measurement. The uncertainty reflects on the laws that express the correlations between physical quantities, and then on the consequent scientific theories.

The reliability of measurements determines the limits of validity of scientific theories, as well as of their technological applications.

2 Measurement Units

Physical quantities are organized in systems of units. This chapter is mainly dedicated to the International System (SI), but other systems, used in specialized fields of physics, are mentioned as well. The chapter ends with an introduction to the dimensions of physical quantities and to the main applications of dimensional analysis.

2.1 Base and Derived Quantities

In Sect. 1.3, it has been shown that every measurement is based on the possibility that some quantities (the additive quantities) can be directly measured by comparison with a unit standard $\mathcal{U}$.

To describe objects and phenomena of the physical world, many quantities are used, both additive and nonadditive, connected by analytical relations. In principle, one could choose an arbitrary unit standard for every additive quantity. This choice would, however, lead to the introduction of inconvenient proportionality factors, and would require us to define and to maintain a large number of measurement standards.

Example 2.1. Let us consider three quantities, length ℓ, time t, and velocity v, which are temporarily labeled $\mathcal{G}_\ell$, $\mathcal{G}_t$, and $\mathcal{G}_v$, respectively. For a uniform motion, the velocity is defined as $\mathcal{G}_v = \Delta\mathcal{G}_\ell/\Delta\mathcal{G}_t$. In principle, one can independently choose the unit standards $\mathcal{U}$ of the three quantities. Possible choices could be the Earth radius $\mathcal{U}_\ell$, the period of the Earth rotation $\mathcal{U}_t$, and the velocity tangential to the equator $\mathcal{U}_v$, respectively. Let us now consider the values X of the quantities $\mathcal{G}$, as defined by (1.1): $\mathcal{G} = X\mathcal{U}$. Velocity is, by definition, the ratio between space and time. According to the independent choice of unit standards, the value of the velocity would be connected to the space and time units through the relation $X_v = (1/2\pi)\,X_\ell/X_t$. In fact, a point fixed at the equator moves with unit velocity $\mathcal{G}_v = 1\,\mathcal{U}_v$, traveling the distance $\mathcal{G}_\ell = 2\pi\mathcal{U}_\ell$ in the unit time $\mathcal{G}_t = 1\,\mathcal{U}_t$.

In this example, the velocity unit is connected to the space and time units through the $(1/2\pi)$ factor. To avoid, or at least to reduce the number of factors different from unity in the relations connecting the units of different physical quantities, it is convenient to arbitrarily choose the unit standards of only a very small number of quantities. For the other quantities, the units are univocally defined through analytical relations.

Example 2.2. Let us choose as arbitrary unit standards the meter (m) for lengths ℓ and the second (s) for times t; the unit of velocity v is then the meter per second (m/s), defined by the relation $v = \ell/t$.

One distinguishes:

(a) *Base quantities*, for which the unit standard is arbitrarily defined
(b) *Derived quantities*, for which the unit standard is defined through analytical relations that connect them to the base quantities

Establishing a *system of measurement units* consists of:

(a) Choosing a particular partition between base and derived physical quantities
(b) Defining the unit standards of the base quantities

The first attempt at establishing a system of units was made by the French revolutionary government in 1790, and led to the proposal of the Decimal Metric System in 1795. Afterwards, various other systems of units have been introduced, many of which are still in use. The increasing demand of standardization, connected to the development of trade and scientific research, led, since 1895 (Meter Convention), to various international agreements on unit systems. In the last decades, there has been a convergence towards the *International System* (SI), which is treated in Sect. 2.3.

A system of units is said to be

- *Complete,* when all physical quantities can be deduced from the base quantities through analytical relations
- *Coherent,* when the analytical relations defining the derived units do not contain proportionality factors different from unity
- *Decimal,* when all multiples and submultiples of units are powers of ten

2.2 Measurement Standards

The unit standards of the fundamental quantities are physically realized by means of *measurement standards*. There are standards also for many derived quantities. The main properties characterizing a measurement standard are:

(a) *Precision*
(b) *Invariability* in time
(c) *Accessibility*, say the possibility that everyone can have access to the standard

(d) *Reproducibility*, say the possibility of reproducing the standard in case of destruction

One can distinguish two fundamental kinds of measurement standards, *natural standards* and *artificial standards*. The distinction is clarified by the following example.

Example 2.3. Let us consider the evolution of the length standard. In 1795, the *meter* was for the first time introduced as the fraction $1/10^7$ of the arc of an Earth meridian from a Pole to the equator (natural standard). In 1799, an artificial standard was built, made by a platinum rule (precision $10 \div 20\,\mu m$). In 1889, the old standard was substituted by a rule made of an alloy 90% platinum + 10% iridium (precision $0.2\,\mu m$). In 1960, a natural standard was again introduced, the optical meter, defined as a multiple of the wavelength of the red-orange light emitted by the isotope 86 of krypton (precision $0.01\,\mu m$). Finally, since 1983, the definition of the meter has been based on the product between the speed of light and an interval of time.

Natural standards guarantee reproducibility and invariability, although sometimes at the expenses of accessibility.

The standards of the highest precision are called *primary standards*. More accessible, although less precise, *secondary standards* are periodically calibrated against primary standards. Standards of current use, the *working standards*, are in turn calibrated against secondary standards.

2.3 The International System of Units (SI)

The International System (SI) divides the physical quantities into base quantities and derived quantities, and defines the names and symbols of their units. The SI defines the measurement standards of the base quantities. In addition, the SI gives the rules for writing names and symbols of the units, as well as the names and symbols of the prefixes (see Appendix B.1). The SI is complete, coherent, and decimal (with the exception of time measurements).

Base Units

The SI is founded on seven base quantities. Their names, units and symbols are listed in Table 2.1. The official definitions of the base SI units are listed in Appendix B.1.

Time Interval

The unit of time, the *second*, is defined with reference to the period of the electromagnetic radiation that is emitted by the isotope 133 of Cesium (say the isotope whose nucleus contains 55 protons and 78 neutrons) during the

Table 2.1. Base quantities of the International System, their units and symbols.

Quantity	Unit	Symbol
Time interval	second	s
Length	meter	m
Mass	kilogram	kg
Amount of substance	mole	mol
Temperature	kelvin	K
Electric current	ampere	A
Luminous intensity	candela	cd

transition between two hyperfine energy levels of its ground state. The time standard is thus a natural standard.

The ground state of an atom corresponds to the electronic configuration of minimum energy. The splitting of the ground state into hyperfine levels is due to the interaction of electrons with the nuclear magnetic moment; the difference in energy ΔE between the hyperfine levels is much smaller than the difference between the principal levels. During the transition between the two levels of ^{133}Cs labeled $F = 4$, $M = 0$, and $F = 3$, $M = 0$, respectively, electromagnetic waves are emitted with frequency $\nu = \Delta E/h \simeq 10^{10}$ Hz, say in the microwave region (h is the Planck constant). The second is defined as 9 192 631 770 times the period $T = 1/\nu$. The primary standard of time is realized by the cesium clock, whose maximum relative uncertainty is 1×10^{-12}, corresponding to 1 µs every twelve days.

Length

The unit of length, the *meter*, is defined as the distance covered by light in vacuum in 1/299 792 458 seconds. As a consequence of this definition, the exact value $c = 299\,792\,458$ m/s has been attributed, since 1983, to the velocity of light in vacuum, one of the fundamental constants of physics.

Mass

The unit of mass, the *kilogram*, is the mass of a cylindric body made by an alloy of platinum and iridium, that is preserved in Sèvres (France). This is the only artificial standard of the SI. Its relative precision is of the order of 10^{-9}.

Amount of Substance

The *mole* is the amount of substance of a system containing as many elementary entities (atoms, molecules, electrons, etc.) as there are atoms in 0.012 kg of isotope 12 of carbon (say the most abundant isotope of carbon, whose nucleus contains six protons and six neutrons). The number N_0 of elementary entities within a mole is called the *Avogadro number*; its approximate value is $N_0 \simeq 6.022 \times 10^{23}$.

Temperature

The *kelvin* is the fraction 1/273.16 of the thermodynamic temperature of the triple point of water. The triple point of water, say the thermodynamic state of equilibrium of the solid, liquid, and gas phases (Fig. 2.1, left), corresponds to a temperature of 273.16 K and a pressure of 610 Pa. The relative precision of the kelvin is 1×10^{-6}.

Temperature is a nonadditive quantity. The absolute thermodynamic temperature is defined in relation to the efficiency of the Carnot cycle; its measurement corresponds to the measurement of the ratio between two additive quantities, for example, two heat quantities.

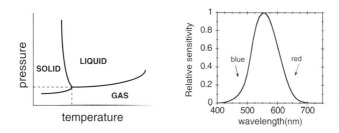

Fig. 2.1. Left: schematic representation of the phase diagram of water. Right: sensitivity curve of the human eye as a function of light wavelength.

Electric Current Intensity

The definition of the unit of electric current intensity, the *ampere*, refers to the force F per unit length ℓ between two parallel conductors placed at a distance d and carrying the same current I:

$$F/\ell = 2k_m I^2/d \; .$$

A numerical value 10^{-7} is attributed to the constant k_m. One ampere corresponds to the current that produces a force of 2×10^{-7} N per meter of length.

For the practical realization of standards, one prefers to rely on the Ohm law $I = V/R$, and obtain the ampere as a ratio between the units of potential difference (volt) and resistance (ohm). The standards of volt and ohm are realized by means of two quantum phenomena, the Josephson effect and the quantum Hall effect, respectively.

Luminous Intensity

Luminous intensity is the base quantity of photometry, say the discipline dealing with the sensitivity of the human eye to electromagnetic radiation.

The luminous intensity corresponds to the flux of energy radiated by a source within the unit solid angle, weighted by the average sensitivity of the human eye (Fig. 2.1, right). Photometric measurements are relevant in the fields of astronomy, photography, and lighting. The unit of luminous intensity, the *candela*, corresponds to the intensity, in a given direction, of a source that emits monochromatic radiation of frequency 540×10^{12} hertz and that has a radiant intensity of 1/683 watt per steradian.

Derived Units

The derived quantities are defined in terms of the seven base quantities via a system of simple equations. There are no conversion factors different from one (the SI is coherent). Some derived units have special names and are listed in Appendix B.1. Here are some relevant examples.

Example 2.4. Acceleration is a derived quantity. By definition, it is the derivative of velocity with respect to time. Its unit is the ratio between the unit of length and the square of the unit of time, say $1\,\mathrm{m\,s^{-2}}$. The unit of acceleration has no special name.

Example 2.5. Plane angle and *solid angle* are derived quantities. Their units have special names, *radian* and *steradian*, respectively. The radian (rad) is the plane angle that subtends, on a circumference centered on its vertex, an arc whose length is equal to the radius. The steradian (sr) is the solid angle that subtends, on a sphere centered on its vertex, a spherical cap whose area is equal to the square of the radius.

Example 2.6. Force F is a derived quantity. By the fundamental law of dynamics, $F = ma$, the unit of force is referred to the units of mass and acceleration. The unit of force has a special name, the *newton* (N), and is defined as $1\,\mathrm{N} = 1\,\mathrm{Kg\,m\,s^{-2}}$.

2.4 Other Systems of Units

In spite of the SI being an internationally adopted complete system, several other systems are still in use. We consider here the systems most relevant for physics.

cgs Systems

In cgs systems, the fundamental mechanical quantities are length, mass, and time, as in the SI. The corresponding units are centimeter, gram, and second (whence the acronym cgs). The differences between cgs and SI systems are limited, for mechanical quantities, to multiplicative factors, powers of 10, in the values of base and derived quantities, and to the name of units.

2.4 Other Systems of Units

The substantial difference between cgs and SI systems concerns the electromagnetic quantities. While the SI has a base quantity for electromagnetism (the electric current intensity), in csg systems all electromagnetic units are derived from mechanical units. Several different cgs systems exist, depending on the relation that is used to define electromagnetic units as a function of mechanical units.

The *electrostatic cgs system* defines the electric charge unit (statcoulomb) through the Coulomb law

$$F_e = K_e \, q_1 \, q_2 / r^2 \, , \tag{2.1}$$

by imposing $K_e = 1$, dimensionless.

The *electromagnetic cgs system* defines the current unit (abampere) through the law of electrodynamic interaction between currents

$$F_m = 2 \, K_m \, I_1 \, I_2 \, \ell / d \, , \tag{2.2}$$

by imposing $K_m = 1$, dimensionless.

The *Gauss symmetrized cgs system* uses the electrostatic cgs units for electrical quantities and the electromagnetic cgs units for magnetic quantities. The symmetrized cgs system is still frequently used in theoretical physics.

Practical Systems

In the past, many practical units have been introduced in different fields of science and technology. After the introduction of the SI, practical units should not be used. Several exceptions limited to specialistic fields are, however, still accepted. Let us list here some non-SI units that are frequently used in physics; other examples can be found in Appendix B.1.

The *atomic mass unit* (u) is 1/12 of the mass of an atom of carbon 12, say the isotope whose nucleus contains six protons and six neutrons. The approximate value is $1 \, \text{u} \simeq 1.66 \times 10^{-27}$ kg.

The *electronvolt* (eV) is the energy gained by an electron when crossing an electric potential difference of 1 V. The approximate value is $1 \, \text{eV} \simeq 1.602 \times 10^{-19}$ J.

The *astronomic unit* (au), roughly corresponding to the distance between the Earth and Sun, is used for expressing distances within the solar system. Its approximate value is $1 \, \text{au} \simeq 1.496 \times 10^{11}$ m.

Plane angles are often measured using the *degree* (°) and its nondecimal submultiples: the minute, $1' = (1/60)°$, and the second, $1'' = (1/3600)°$.

The *ångström* (Å) is often used to measure distances at the atomic level: $1 \, \text{Å} = 0.1 \, \text{nm} = 10^{-10}$ m.

British Systems

In some countries, such as Great Britain (UK) and the United States of America (USA), some British units are still in use (a partial list can be found in Appendix B.3).

British systems are not decimal. For example, the base unit of length is the *inch*, and its most important multiples are the *foot*, corresponding to 12 inches, and the *yard*, corresponding to three feet.

Some units have different values in the UK and in the USA. For example, the gallon, unit of volume, corresponds to $4.546\,\mathrm{dm}^3$ in the UK and $3.785\,\mathrm{dm}^3$ in the USA.

Natural Systems

In some specialistic fields of physics, it is customary to use peculiar units, which are called *natural* because they refer to particularly relevant fundamental quantities.

The *Hartree atomic system* is often used when describing phenomena at the atomic level. Its base quantities are:

- *Mass:* the unit is the electron rest mass m_e (in SI, $m_e \simeq 9.109 \times 10^{-31}$ kg)
- *Electric charge:* the unit is the electron charge e (in SI, $e \simeq 1.602 \times 10^{-19}$ C)
- *Action* (product of energy and time): the unit is the quantum of action h (Planck constant) divided by 2π, $\hbar = h/2\pi$ (in SI, $\hbar \simeq 1.054 \times 10^{-34}$ J s)

The *Dirac system* is often used in elementary particle physics. Its base quantities are:

- *Mass:* the unit is again the electron rest mass, m_e
- *Velocity:* the unit is the velocity of light in vacuum, c (in SI, $c = 299\,792\,458\,\mathrm{m\,s^{-1}}$)
- *Action:* the unit is again $\hbar = h/2\pi$

2.5 Dimensional Analysis

The choice of the base units reflects on the numerical values of both base and derived quantities. Dimensional analysis deals with this topic, and represents a useful tool for testing the soundness of equations connecting physical quantities.

Dimensions of Physical Quantities

Let us suppose that the unit of length, the meter, is substituted by a new unit, L times smaller; for example, the centimeter is $L = 100$ times smaller. As a consequence of the new choice of the length unit, the values

- of length are multiplied by $\quad$ L
- of time are multiplied by $\quad L^0 = 1$
- of volume are multiplied by $\quad L^3$
- of velocity are multiplied by $\quad$ L

The exponent of L is the *dimension* with respect to length:

- Length has dimension $\quad$ 1 $\quad$ with respect to length
- Time has dimension $\quad$ 0 $\quad$ with respect to length
- Volume has dimension $\quad$ 3 $\quad$ with respect to length
- Velocity has dimension $\quad$ 1 $\quad$ with respect to length

The dependence of the value X of whichever base or derived quantity on the units of the base quantities $A, B, C, \ldots$ is symbolically expressed by a *dimensional equation*:

$$[X] = [A]^\alpha \ [B]^\beta \ [C]^\gamma \ldots \tag{2.3}$$

where α, β, γ are the dimensions of X with respect to A, B, and C, respectively.

Dimensional analysis is mainly used in mechanics: here we consider only the dimensions with respect to length, mass, and time, which are symbolized by L, M, T, respectively. For example, the dimensions of velocity are expressed by the equation

$$[v] = [L]^1 \ [T]^{-1} \ [M]^0 \ , \tag{2.4}$$

and the dimensions of work and energy are expressed by the equation

$$[W] = [E] = [L]^2 \ [T]^{-2} \ [M]^1 \ . \tag{2.5}$$

Quantities characterized by the same dimensions are said to be dimensionally homogeneous.

Dimensionless Quantities

Some quantities have zero dimension with respect to all base quantities:

$$[L]^0 \ [T]^0 \ [M]^0 \ . \tag{2.6}$$

It is the case of pure numbers (3, $\sqrt{2}$, π, ...) and of dimensionless quantities, say quantities defined by the ratio between two homogeneous quantities. The value of dimensionless quantities does not depend on the particular choice of the base units.

Example 2.7. Plane angles are dimensionless quantities; their value measured in radians is the ratio between the length of the arc and the length of the radius. Also solid angles are dimensionless quantities; their value measured in steradians is the ratio between two squared lengths.

Example 2.8. The absolute density of a substance is the ratio between its mass and its volume: $\rho = m/V$. The dimensions of the absolute density are given by $[\rho] = [L]^{-3}[T]^0[M]^1$. The relative density of a substance is the ratio between its absolute density and the absolute density of water at the temperature of $4°C$. The relative density is a dimensionless quantity.

Principle of Dimensional Homogeneity

The usefulness of dimensional analysis is founded on the principle of dimensional homogeneity, stating that it is possible to sum or equate only dimensionally homogeneous quantities. Otherwise stated, any equation between physical quantities

$$A + B + C + \ldots = M + N + P + \ldots \tag{2.7}$$

is true only if $A, B, C, \ldots, M, N, P, \ldots$ are dimensionally homogeneous monomials. In particular, transcendental functions (sin, cos, exp, log, etc.) are dimensionless, and their arguments must be dimensionless.

Applications of Dimensional Analysis

Let us consider here the most important applications of dimensional analysis.

Test of Equations

The dimensional homogeneity is a *necessary condition* for the correctness of equations connecting physical quantities, such as (2.7). This means that, for the equation to be true, all terms must have the same dimensions. Dimensional homogeneity is thus the first test of validity of any theoretical relationship.

Example 2.9. Let us consider the oscillations of a mass suspended from a spring. The position x of the mass depends on time t according to a sinusoidal law. The dependence of x on t cannot be expressed by $x = \sin t$, because: (a) the argument t of the sine function is not dimensionless; (b) the sine function is dimensionless, and x has the dimension of length. A valid expression is $x = A\sin(\omega t)$, where A is a constant with the dimension of length, and ω is a constant with the dimension of an inverse time.

The dimensional homogeneity, however, is not a *sufficient condition* for the correctness of an equation, since:

1. Dimensional analysis cannot test the correctness of numerical values

2. There exist some quantities that, in spite of being dimensionally homogeneous, have a completely different meaning (e.g., mechanical work and the momentum of a force).

Example 2.10. We want to determine the trajectory of a missile that is thrown at an angle θ with respect to the horizontal, with initial velocity v_0. After some kinematical calculations, we find the following equation,

$$z = -\frac{g}{2 v_0 \cos^2 \theta} x^2 + x \cos \theta, \qquad \text{[wrong!]} \qquad (2.8)$$

where x and z are the horizontal and vertical coordinates, respectively. A dimensional check shows that (2.8) is wrong. We repeat the calculations and find

$$z = -\frac{g}{2 v_0^2 \cos^2 \theta} x^2 + x \cos \theta; \qquad \text{[wrong!]} \qquad (2.9)$$

the equation is now dimensionally homogeneous; in spite of this, it is still wrong. The right equation is

$$z = -\frac{g}{2 v_0^2 \cos^2 \theta} x^2 + x \tan \theta. \qquad (2.10)$$

The exchange of $\tan \theta$ by $\cos \theta$ in the last term of (2.9) led to a dimensionally homogeneous wrong equation.

In order to use dimensional analysis for testing equations, it is necessary to perform calculations in literal form. Numerical values should be inserted only at the end of the test.

Deduction of Physical Laws

In particular cases, dimensional analysis allows one to find analytical relations between different quantities characterizing a physical phenomenon. Obviously, no information on numerical values can be obtained, and dimensionless constants are neglected.

Example 2.11. The period T of a pendulum can depend, in principle, on the mass m and length ℓ of the pendulum, on the acceleration of gravity g, and on the amplitude θ_0 of the angular oscillation. The dimensional dependence of T on m, ℓ, and g can be expressed as

$$[T] = [m]^\alpha [\ell]^\beta [g]^\gamma, \qquad (2.11)$$

say, taking into account the dimensions of m, ℓ, and g,

$$[T] = [M]^\alpha [L]^{\beta+\gamma} [T]^{-2\gamma}. \qquad (2.12)$$

The principle of dimensional homogeneity demands that

$$\alpha = 0, \quad \beta + \gamma = 0, \quad \gamma = -1/2, \tag{2.13}$$

whence

$$T = C\sqrt{\ell/g}, \tag{2.14}$$

where C is a dimensionless constant. It is important to note that one cannot determine, by dimensional analysis, the possible dependence of the period on the amplitude θ_0 (dimensionless), nor the value of the constant C.

Actually, the period of a pendulum can easily be determined, including the value of the dimensionless constant C and the dependence on the amplitude θ_0, by properly solving the equation of motion. The foregoing application of dimensional analysis to the determination of the period of a pendulum has thus mainly didactical interest.

In the case of very complex physical systems, however, for which a complete theory does not exist or is exceedingly complicated (such as in some problems of fluid dynamics), dimensional analysis can represent a very useful tool.

Physical Similitude

Large complex systems are often studied with the help of models in reduced scale (e.g., in applications of hydraulic and aeronautic engineering, elasticity, and heat transmission). A basic problem is to evaluate how the scale reduction affects the different properties of the model with respect to the real system. For example, a reduction of linear dimensions by a factor $L = 100$ produces a reduction of volumes by a factor 10^6, which corresponds to a reduction of masses by the same factor 10^6, if the densities are the same in the model and in the real system.

Dimensional analysis is of great help in this task, because it accounts for the influence of the reduction of base units on the values of derived quantities. In particular, dimensionless quantities, such as relative density or more specialized quantities (e.g., the Reynolds number or the Mach number) are very effective, because their values do not depend on scale factors.

Problems

2.1. The unit of mechanical work is the joule in the International System, and the erg in the cgs systems. Calculate the conversion factor between joule and erg.

2.2. The measure of a plane angle in degrees is $\theta = 25°\,7'\,36''$. Express the measure of the angle in radians.

2.3. The mass of an iron atom is 55.847 atomic mass units (average value of the masses of the different isotopes, weighted by their natural abundance). Calculate the mass, in kilograms, of three moles of iron atoms.

2.4. The interaction between the two atoms of a hydrogen molecule H_2 can be assimilated, to a good approximation, to an elastic harmonic spring. The potential energy is $E_p(x) = kx^2/2$, where $x = r - r_0$ is the deviation of the instantaneous distance r from the equilibrium distance $r_0 = 0.74$ Å, and k is the elastic constant, whose value is $k = 36\,\text{eV}/\text{Å}^2$.

In classical mechanics, the elastic constant of a spring is the ratio between force and deformation, $k = -F_e/x$. Express the elastic force k of the hydrogen molecule in the SI unit, newton/meter.

The frequency of oscillation is $\nu = (k/\mu)^{1/2}/2\pi$, where $\mu = 2m$ is the reduced mass of the molecule, and m is the mass of the hydrogen atom. Taking into account that a hydrogen atom contains one proton and one electron, calculate its mass m (using Table C.2 of Appendix C.2), and verify that the frequency of oscillation is $\nu = 1.3 \times 10^{14}$ Hz.

2.5. *Dimensional analysis.* The gravitational interaction between two masses m_1 and m_2 gives rise to an attractive force, proportional to the product of the masses and inversely proportional to the square of their distance: $F \propto m_1 m_2/r^2$. Verify that the equation $F = m_1 m_2/r^2$ is dimensionally wrong.

The right expression is $F = G m_1 m_2/r^2$, where G is the gravitational constant; determine the dimensions of G.

3 Measuring Instruments

The measurement of physical quantities is done by means of measuring instruments. In the everyday practice of scientific research or technological applications, many different instruments are used. The instruments are generally produced in series, but sometimes they are expressly built for specific purposes. This chapter is dedicated to the introduction of some general criteria for classifying instruments and evaluating their performance.

3.1 Functional Elements

In principle, the operation of an instrument can be schematized as in Fig. 3.1: the quantity $\mathcal{G}$ is compared with the unit standard $\mathcal{U}$; the result of the measurement, say the measure X, is often transformed into the value Z of an easily readable output quantity $\mathcal{M}$ (such as the displacement of an index on a dial). In Fig. 3.1, the fourth arrow (E) represents energy, which often has to be fed to the instrument.

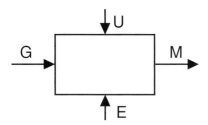

Fig. 3.1. Schematic representation of a measuring instrument: $\mathcal{G}$ and $\mathcal{M}$ are the input and output quantities, respectively, $\mathcal{U}$ is the standard unit, and E is energy.

A few instruments have a simple logical structure, where the quantity $\mathcal{G}$ is directly compared with the standard $\mathcal{U}$.

Example 3.1. A ruler for length measurements. The input quantity $\mathcal{G}$ is the length to be measured. The unit standard $\mathcal{U}$ (typically a millimeter) and its multiples are engraved on the ruler. The measure $X(\mathcal{G})$ is directly readable on the rule.

In the great majority of instruments, the logical structure is more complex: the comparison with the unit standard is made through a *calibration*, which is generally performed by the manufacturer. The quantity $\mathcal{G}$ can undergo various manipulations and transformations to other quantities.

Example 3.2. The mercury-in-glass thermometer. The input quantity $\mathcal{G}$ is the temperature of the external environment. The output quantity $\mathcal{M}$ is the height of the mercury column. The instrument is calibrated in such a way that the values of the input quantity $X(\mathcal{G})$ can be directly read on the scale.

Fig. 3.2. Schematic representation of an instrument as a chain of functional elements. $\mathcal{G}$ and $\mathcal{M}$ are the input and output quantities, respectively.

It can often be convenient to represent the instrument as a *measuring chain*, decomposing its logical structure into a sequence of functional elements, each one devoted to a well-defined task (Fig. 3.2).

Input Element (Sensor)

The sensor is the first element of the measurement chain, and is directly affected by the input quantity $\mathcal{G}$.

Examples of sensors are the bulb of a mercury-in-glass thermometer, the two-metal junction of a thermocouple thermometer, and the pair of terminals of an instrument for measuring electrical currents or potential differences.

Output Element

The last element of the measurement chain conveys the value $Z(\mathcal{M})$ of the output quantity, which gives information on the value $X(\mathcal{G})$ of the input quantity. The output element can be directly readable by an operator, such as an index on a dial, a digital display, the pen of a plotter, or a printing device. Alternatively, the output element can produce signals, typically electrical, suitable to be fed to another mechanical or electrical instrument, such as an actuator or a computer.

Example 3.3. An ambient thermostat measures the temperature of a room and compares it with a preset value. The measurement result is sent, as an electric signal, to an actuator or a computer, which controls a pump or a valve of the air conditioning system.

Intermediate Elements: Transducers, Amplifiers, Manipulators

Within a measuring chain, the measured quantity can undergo various transformations to other quantities, which can be more easily manipulated, transmitted, or displayed. Frequently, mechanical quantities are transformed into electrical quantities. The transforming element is called a *transducer*. The value of a quantity can also be amplified, or undergo mathematical operations, such as addition, integration, and so on. The flexibility of electrical signals for this kind of manipulation is one of the reasons for converting mechanical or thermal quantities into electrical quantities.

Example 3.4. In a resistance thermometer, the sensor is an electrical resistor carrying a constant current. A variation of temperature induces a variation of electrical resistance, which in turn induces a variation of potential difference. The weak variations of potential difference are amplified in order to drive the displacement of an index on a calibrated dial.

Example 3.5. An electronic scale is designed to measure masses m. To this purpose, the scale transforms the weight mg into an electrical quantity, a current or a potential difference. The variations of the electrical quantity are amplified in order to drive a calibrated display.

It is worth noting that the logical decomposition of an instrument into a chain of elements does not necessarily correspond to a real physical situation. Frequently, a single physical component performs the logical functions of two or more functional elements.

3.2 Classifications of Instruments

Different classifications of instruments are possible, in relation to their operating properties.

Absolute and Differential Instruments

In *absolute* instruments, a variation of the value X of the input quantity $\mathcal{G}$ is transformed into a corresponding variation of the value Z of the output quantity $\mathcal{M}$ (such as the deviation of an index on a dial). The measuring chain is open (Fig. 3.3, left). Examples of absolute instruments are the spring dynamometer and the mercury-in-glass thermometer.

In *differential* instruments, the unknown value X of the input quantity $\mathcal{G}$ is compared with a known value Y of the same quantity. In some cases, it is the difference $X - Y$ which is of interest, and is output from the instrument. In other cases, the known value Y is varied until the difference $X - Y$ is zero. The output element of the measuring chain is then a zero detector (e.g., a dial centered on zero). In this second type of instrument, the measuring chain is closed, and is characterized by a feedback element, typically a manual or automatic actuator that varies the known value Y (Fig. 3.3, right). The instruments with the output centered on zero are generally more accurate than the absolute instruments, but less suitable for measuring quantities varying in time.

Example 3.6. The analytical equal-arms balance is a simple example of a differential instrument. The measurement of an unknown mass m on one scale-pan requires us to manually change the number of standard masses on the other scale-pan, until mechanical equilibrium is achieved.

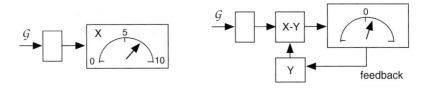

Fig. 3.3. Schematic comparison of an open chain instrument (left) and a closed chain instrument (right).

Analog and Digital Instruments

In *analog instruments*, the input value X is transformed, at the end of the measuring chain, into an analog signal, which can assume a continuous range of values, such as the deviation of an index or the value of an electric current (Fig. 3.4, left).

In *digital instruments*, the input value X is transformed, at the end of the measuring chain, into a number, directly readable on a display (Fig. 3.4, right), or codified as an electric signal, suitable as input to a computer. The transformation of the analog input signal into a digital output signal is generally obtained by means of electronic devices, called *analog to digital converters* (ADC).

Fig. 3.4. Displays of an analog instrument (left) and of a digital instrument (right).

Displaying and Recording Instruments

In *displaying instruments*, the output value is available to the observer only during the measurement. Examples are the analytical equal-beam balance and the mercury-in-glass thermometer.

In *recording instruments*, the output value is stored, in analog or digital form, on suitable supports, such as paper, magnetic disks, semiconductor memories, and so on.

Active and Passive Instruments

Passive instruments get the energy necessary to their working directly from the system on which the measurement is done.

Active instruments are instead fed from an energy source external to the system under measurement (such as a battery).

Example 3.7. A mercury-in-glass thermometer is a passive instrument, a resistance thermometer is an active instrument.

3.3 Static Characteristics of Instruments

In Sects. 3.3 and 3.4, attention is focused on the performance of instruments for the measurement of physical quantities that are constant in time. The dynamical properties of instruments, relative to measurement of quantities variable in time, is considered in Sect. 3.5.

Measurement Range

The measurement range is the interval of values X of the input quantity $\mathcal{G}$ within which the instrument operates within a specified degree of accuracy. The measurement range is included between a lower limit and an upper limit. The *upper safety limit* is the maximum value of the input quantity that can be supplied without damaging the instrument. In some instruments, the range can be varied by means of suitable selectors, such as rotating knobs.

Example 3.8. In the mercury-in-glass thermometer, the measurement range is defined by the minimum and maximum values on the scale (e.g., $-10°C$ and $+100°C$). The upper limit generally is also the safety limit, higher temperatures can cause the breakdown of the thermometer.

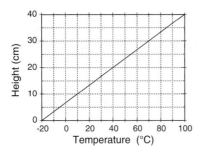 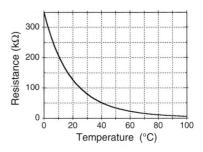

Fig. 3.5. In the mercury-in-glass thermometer, the relation between temperature and height of the mercury column is linear (left). In a semiconductor thermometer, the relation between temperature and electrical resistance is nonlinear (right).

Linearity

An instrument is said to be linear if the relation connecting the values Z of the output quantity $\mathcal{M}$ to the values X of the input quantity $\mathcal{G}$ is linear.

Example 3.9. The mercury-in-glass thermometer is, to a good approximation, a linear instrument. Within the usual measurement range, the thermal expansion coefficient of mercury is constant, and the height of the mercury column linearly increases with temperature (Fig. 3.5, left).

Example 3.10. A thermometer with a semiconductor probe (thermistor) is a good example of a nonlinear instrument. The input quantity is temperature, the output quantity is the electrical resistance of the thermistor. When temperature increases, the thermistor resistance decreases such as in Fig. 3.5, right.

Sensitivity

The sensitivity is the ratio $\Delta Z/\Delta X$ between the variations of the values of the output and input quantities.

Example 3.11. In a mercury-in-glass thermometer, the input quantity is temperature, and the output quantity is the height of the mercury column. In Fig. 3.6, thermometer A has a sensitivity five times higher than thermometer B, because the same variation of temperature gives rise to a variation of the mercury column height five times larger in thermometer A than in thermometer B.

In *linear instruments*, the sensitivity $\Delta Z/\Delta X$ is constant over the full measurement range.

Example 3.12. The thermometer on the left of Fig. 3.5 has a constant sensitivity $\Delta h/\Delta \theta = 0.33$ cm/°C.

3.3 Static Characteristics of Instruments   33

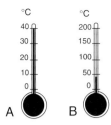

Fig. 3.6. Two mercury-in-glass thermometers with different sensitivities.

In *nonlinear instruments*, the sensitivity varies within the measurement range, such as the slope of the curve $Z(X)$. The sensitivity is thus defined as the first derivative dZ/dX.

Example 3.13. In the semiconductor thermometer of Fig. 3.5 (right), the absolute value $|dR/d\theta|$ of sensitivity varies from 18.4 kΩ/°C at 0°C to 0.18 kΩ/°C at 100°C.

In *analog instruments*, the display generally consists of a graduated scale. The sensitivity $\Delta Z/\Delta X$ is often expressed taking the number of divisions on the scale as ΔZ. In this case, sensitivity is the inverse of resolution (see below).

Example 3.14. A ruler with resolution $\Delta X = 1$ mm has a sensitivity $1/\Delta X$.

In *digital instruments*, the sensitivity is evaluated by considering the analog quantity $\mathcal{M}$ immediately before the analog/digital converter.

The *sensitivity threshold* is the smallest value of the input quantity $\mathcal{G}$ that can induce a variation of the output quantity $\mathcal{M}$.

Display Resolution

The display resolution is the smallest variation ΔX of the input quantity $\mathcal{G}$ that can be measured, corresponding to the smallest variation ΔZ of the output quantity $\mathcal{M}$ that can be detected.

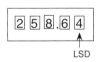

Fig. 3.7. Display resolution of an analog instrument (left) and of a digital instrument (right).

In *analog instruments*, the display resolution generally corresponds to the minimum distance between two ticks of the graduated scale (Fig. 3.7, left).

In *digital instruments*, the display resolution corresponds to the unit value of the least significant digit (LSD) (Fig. 3.7, right).

Example 3.15. The display resolution of common rulers generally is $\Delta X = 1\,\text{mm}$. The display resolution of micrometer gauges generally is $\Delta X = 0.01\,\text{mm}$.

Sensitivity and display resolution are strictly correlated properties. To reduce the display resolution of an instrument, it is generally necessary to increase its sensitivity. In analog instruments, the sensitivity is sometimes expressed as the inverse of resolution, $1/\Delta X$.

There is still no complete agreement on nomenclature. The term *resolving power* is frequently used for the inverse of resolution, $1/\Delta X$, but sometimes also to express the ratio between measurement range and resolution. Also common, although incorrect and misleading, is the use of *resolution* for *resolving power*. The actual meaning of terms is generally clarified by the context.

Transparency

In general, an instrument perturbs the system under measurement. As a consequence, the value X of the input quantity $\mathcal{G}$ is modified by the instrument. One speaks of *transparency* to qualify the degree of this disturbance.

Example 3.16. Measuring a mass with an equal-arms balance does not alter the value of the mass. The equal-arms balance is a transparent instrument.

Example 3.17. Measuring the temperature of a system requires a heat exchange between the system and the thermometer, which intrinsically alters the thermodynamic state of the system, and hence its temperature. The thermometer is not a transparent instrument.

In Sect. 3.2, active and passive instruments have been distinguished. Passive instruments perturb the systems subject to measurement, by extracting the energy necessary for their working. One should, however, notice that not even the perturbation induced by active instruments can be completely eliminated, although, in macroscopic measurements, it can often be reduced below acceptable levels. For measurements on systems at atomic or subatomic scales, the perturbation induced by measuring instruments and procedures is never negligible.

Operating Conditions

The performance of an instrument depends on the environmental conditions. In addition to the quantity $\mathcal{G}$ to be measured, other *influence quantities* can contribute to modify the measurement result: temperature, pressure, humidity, mechanical vibrations, acceleration, electromagnetic fields, and the like.

The *operating conditions* define the intervals of the influence quantities within which the instrument can perform the measurements within a specified degree of accuracy.

Example 3.18. From the technical characteristics of an electronic balance:

Operating temperature	$0 \div 40°C$
Under/over sea-level	$-3400\,\text{m} \cdots +6000\,\text{m}$
Relative air humidity	$15\% \div 85\%$
Vibrations	$0.3\,\text{m/s}^2$

Performing measurements outside the specified operating conditions introduces errors in measured values and can damage the instrument.

3.4 Accuracy of an Instrument

It has been pointed out in Sect. 1.3 that the result of the measurement of a physical quantity is never a single value X. In fact, any instrument is characterized by a finite display resolution ΔX, and cannot give information on variations smaller than ΔX. As a consequence, the result of a single measurement is represented by a continuous interval of values, of width ΔX. By convention, the measure of a physical quantity is generally quoted as

$$X = X_0 \pm \delta X, \tag{3.1}$$

where X_0 is a central value giving the position of the interval on the X-axis and δX is the uncertainty due to the display resolution.

For the time being, the uncertainty is expressed as $\delta X = \Delta X/2$. Later on, in Chap. 4, after a thorough investigation on the origin of uncertainty, it will appear more convenient to express δX in a slightly different way.

In digital instruments, the central value X_0 of (3.1) is directly read on the display. Resolution ΔX and uncertainty δX are given by the unit value of the least significant digit (LSD) and its half, respectively.

Example 3.19. A time interval τ is measured by means of a digital stopwatch. A value 34.27 s is read on the display. The unit LSD value is 0.01 s. The measure has to be quoted as $\tau = (34.27 \pm 0.005)\,\text{s}$.

In analog instruments, the evaluation of the central value X_0 can depend on the position of an object or an index with respect to the ticks of a graduated scale. Two typical situations, encountered when measuring lengths, are exemplified in Fig. 3.8. In the first case (center of figure) the value X_0 can be directly read in correspondence to the scale tick. In the second case (right of figure), one can attribute to X_0 a value intermediate between those corresponding to the two nearest ticks.

Assuming a resolution ΔX smaller than the distance between two contiguous ticks can be an imprudent choice, although sometimes apparently acceptable. In fact, the display resolution of an instrument is generally reduced by the manufacturer to the minimum value consistent with overall accuracy.

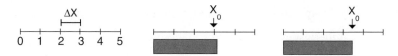

Fig. 3.8. Measurement by means of an analog instrument. Left: the graduated scale, with resolution $\Delta X = 1$ (in arbitrary units). Center: $X = 3 \pm 0.5$. Right: $X = 3.5 \pm 0.5$.

Example 3.20. The height of a paper sheet is measured by a ruler with resolution $\Delta X = 1$ mm. The end of the sheet is equidistant from the ticks corresponding to 296 and 297 mm. It is reasonable to take $X_0 = 296.5$ mm as the central value, but it can be imprudent to assume an uncertainty smaller than half the distance between the ticks, $\delta X = 0.5$ mm

In addition to the finite display resolution, other instrumental factors can influence the result and the quality of a measurement. These factors depend on the instrument structure, the accuracy of its manufacture, the degree of maintenance, the environmental conditions, and so on. Let as give here some examples:

– Calibration defects, such as the limited accuracy of the reference standard and/or of the calibration procedure
– Defects of the zero calibration (e.g., in electronic instruments the zero can drift in time)
– Friction or mechanical plays
– Effects of influence quantities, such as temperature, humidity, and so on

In general, the effects of these different factors on the performance of an instrument are classified in two main categories:

(a) *Systematic errors* are the effects that are always reproduced to the same extent when the measurement is repeated.
(b) *Random fluctuations* or *random errors* are effects that contribute differently and randomly at each repetition of the measurement.

Sometimes, it is also important to take into account the *stability*, say the ability of an instrument to give similar results for measurements of the same quantity repeated over time intervals (typically days or weeks) much longer than the time duration of a single measurement.

The term *accuracy* is used to quote the global performance of an instrument. Accuracy depends on the display resolution, on the influence of systematic and random errors, as well as on long-term stability.

The accuracy of an instrument is generally quoted, in its operation manual, by means of the numerical value of the global uncertainty, say in the form $\pm \delta X$.

When accuracy is not explicitly quoted directly on the instrument or in its operation manual, one can assume that the effects of systematic errors, random fluctuations, and long-term instability are negligible with respect to the resolution ΔX. In such cases, one can quote the global uncertainty as $\delta X = \Delta X / 2$.

Example 3.21. In rulers of common use, with display resolution $\Delta X = 1\,\text{mm}$, accuracy is not quoted. This means that the manufacturer guarantees that the measurement uncertainty due to the instrument cannot be larger than $0.5\,\text{mm}$.

Example 3.22. In the operation manual of a digital thermometer with display resolution $\Delta T = 0.1°\text{C}$, the accuracy is quoted as $\delta T = \pm 0.4°\text{C}$ within the interval from -25 to $+75°\text{C}$. In this case, the overall uncertainty is clearly larger than half the resolution.

In this section, some important concepts concerning the evaluation of instrument accuracy have been introduced. In Chap. 4, it is made clear that the uncertainty does not depend only on instruments, but also on other relevant factors, and the problem of accuracy is accordingly treated within a more general framework.

3.5 Dynamical Behavior of Instruments

In previous Sects. 3.3 and 3.4, the performance of instruments has been considered for the measurement of constant quantities. In this section, an introductory account is given of the behavior of instruments when measuring time-dependent quantities.

Time-Dependent Quantities

Different kinds of time dependences of physical quantities can be distinguished. Three of them, particularly important, are considered in the following examples and in Fig. 3.9.

Example 3.23. Let us consider an oscillating pendulum. The amplitude Φ_0 of the oscillations decreases with time, because of friction. For small oscillations, however, the variation is slow, and Φ_0 can be considered as constant during many periods of oscillation (Fig. 3.9, left).

Example 3.24. A mercury-in-glass thermometer is initially in air at temperature $T = 20°\text{C}$; at the time $t = t_0$, the thermometer is immersed in water at temperature $T = 80°\text{C}$. The environment temperature T thus varies in time according to a step law (Fig. 3.9, center):

$$T(t) = \begin{cases} 20°\text{C for } t < t_0 \\ 80°\text{C for } t \geq t_0 \end{cases}. \tag{3.2}$$

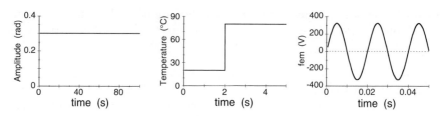

Fig. 3.9. Three particularly important time dependences of physical quantities: constant (left), Example 3.23; step (center), Example 3.24; sinusoidal (right), Example 3.25.

Example 3.25. The electric potential difference in domestic distribution networks varies sinusoidally in time according to the law $V(t) = V_0 \sin(2\pi\nu + \phi)$, where $\nu = 50\,\text{Hz}$ in Europe and $\nu = 60\,\text{Hz}$ in the USA (Fig. 3.9, right).

When a physical quantity is not constant in time, its measurement requires an instrument able to follow its time variations with suitable speed. Knowing the dynamical behavior of an instrument means knowing how the value of the output quantity $Z(t)$ is modified by the variation of the value of the input quantity $X(t)$.

The above examples 3.23–3.25 refer to quite simple behaviors (constant, step, and sinusoidal, respectively). Much more complicated time dependences can be found in the real world. One can, however, demonstrate that periodic or nonperiodic functions can be expressed as a series (Fourier series) or an integral (Fourier integral), respectively, of sinusoidal functions of different frequencies. The relationship connecting the sinusoidal input signal $X(t)$ to the output signal $Z(t)$ as a function of frequency is called the *response function* of an instrument. The knowledge of the response function of an instrument allows one, in principle, to know its dynamic behavior for any time dependence of the measured quantity.

Mathematical Models of Instruments

Mathematical models of instruments are useful to study the relation between input values $X(t)$ and output values $Z(t)$.

The simplest case is represented by *zero-order instruments*, whose mathematical model is a relation of direct proportionality between $X(t)$ and $Z(t)$:

$$a_0 Z = b_0 X , \quad \text{say} \quad Z = \frac{b_0}{a_0} X . \tag{3.3}$$

The proportionality constant $k = b_0/a_0$ corresponds to the *static sensitivity* introduced in Sect. 3.3. According to (3.3), the response $Z(t)$ is instantaneous, independently of the speed of variation of $X(t)$. A zero-order instrument is obviously an ideal model; it is, however, a good approximation for instruments

3.5 Dynamical Behavior of Instruments

whose response is fast with respect to the time variations of the input quantity $X(t)$.

In *first-order instruments*, the relation between $X(t)$ and $Z(t)$ is described by a first-order differential equation:

$$a_1 \frac{dZ}{dt} + a_0 Z = b_0 X \,. \tag{3.4}$$

The presence of the term $a_1(dZ/dt)$ in (3.4) implies that Z cannot instantaneously follow the variations of X. A variation of X initially reflects on the term $a_1(dZ/dt)$; the smaller is the coefficient a_1, the larger is the derivative (dZ/dt), and the faster is the variation of Z.

In *second-order instruments*, the relation between $X(t)$ and $Z(t)$ is described by a second-order differential equation, corresponding to the equation of motion of a damped and forced harmonic oscillator:

$$a_2 \frac{d^2 Z}{dt^2} + a_1 \frac{dZ}{dt} + a_0 Z = b_0 X \,. \tag{3.5}$$

In general, the mathematical model of many instruments consists of a *linear differential equation with constant coefficients*, whose order represents the order of the instrument:

$$a_n \frac{d^n Z}{dt^n} + a_{n-1} \frac{d^{n-1} Z}{dt^{n-1}} + \cdots + a_1 \frac{dZ}{dt} + a_0 Z = b_0 X \,. \tag{3.6}$$

Example 3.26. The mercury-in-glass thermometer can be described, with good approximation, by a first-order model. Let T_{in} and T be the input and output values of temperature, respectively. A difference between the ambient temperature T_{in} and the temperature T of the thermometer bulb induces a heat flux

$$\frac{dQ}{dt} = -k \frac{S}{d} (T - T_{\text{in}}) \,, \tag{3.7}$$

where t is time, k is the thermal conductivity of the bulb glass, S is the surface area of the bulb, and d is the glass thickness. The heat transfer induces a variation of the bulb temperature,

$$dT = dQ/C \,, \tag{3.8}$$

where C is the heat capacity of mercury within the bulb. By eliminating dQ from (3.7) and (3.8), one obtains the first-order differential equation relating T to T_{in}:

$$C \frac{dT}{dt} + k \frac{S}{d} T = k \frac{S}{d} T_{\text{in}} \,. \tag{3.9}$$

An experimental test of the validity of the first-order model for the mercury thermometer is proposed in Experiment E.7 of Appendix E.

Example 3.27. The spring dynamometer, used for measuring forces, can be described by a second-order model. The input quantity is the force F_{in} to be measured, the output quantity is the displacement Z of the spring end. The model is based on the equation of motion $\sum F_i = m\,(\mathrm{d}^2 Z/\mathrm{d}t^2)$. The active forces, in addition to F_{in}, are the elastic force $F_{\text{el}} = -kZ$ and the damping force $F_{\text{fr}} = -\eta(\mathrm{d}Z/\mathrm{d}t)$. The differential equation connecting F_{in} to Z is second-order:

$$m\frac{\mathrm{d}^2 Z}{\mathrm{d}t^2} + \eta\frac{\mathrm{d}Z}{\mathrm{d}t} + kZ = F_{\text{in}} \,. \tag{3.10}$$

First- and Second-Order Instruments

Once the time dependence of the input quantity $X(t)$ is known, the behavior of the output quantity $Z(t)$ is found by integrating the differential equation of the instrument. Attention is focused here on the first- and second-order instruments. Let us first summarize some basics on differential equations; a more detailed treatment is given in Appendix D.1.

Equations (3.4) and (3.5) are inhomogeneous, because the second member $X(t)$ is different from zero. By setting $X(t) = 0$ in (3.4) and (3.5), one obtains the corresponding homogeneous equations.

The general solution $Z(t)$ of a linear differential equation with constant coefficients is the sum of two functions:

$$Z(t) = Z_{\text{tr}}(t) + Z_{\text{st}}(t) \,, \tag{3.11}$$

$Z_{\text{tr}}(t)$ being the general solution of the homogeneous equation (the index "tr" means "transient")

$Z_{\text{st}}(t)$ being one particular solution of the in-homogeneous equation (the index "st" means "stationary")

The *transient functions* $Z_{\text{tr}}(t)$, solutions of homogeneous equations, only depend on the instrument, not on the input quantity $X(t)$.

For the first-order homogeneous equation,

$$\frac{\mathrm{d}Z}{\mathrm{d}t} + \gamma Z = 0 \,, \quad \text{where } \gamma = a_0/a_1 \,, \tag{3.12}$$

one can easily verify that

$$Z_{\text{tr}}(t) = Z_0\,\mathrm{e}^{-\gamma t} \,. \tag{3.13}$$

The second-order homogeneous equation can be written in the standard form of the damped harmonic oscillator:

$$\frac{\mathrm{d}^2 Z}{\mathrm{d}t^2} + 2\gamma\frac{\mathrm{d}Z}{\mathrm{d}t} + \omega_0^2 = 0\,, \quad (2\gamma = a_1/a_2;\ \omega_0^2 = a_0/a_2)\,, \tag{3.14}$$

where ω_0 is the proper angular frequency, and γ is the damping factor. One can have three different solutions, depending on the relation between γ and ω_0:

$$\gamma < \omega_0 \Rightarrow Z_{tr}(t) = Z_0\, e^{-\gamma t}\sin(\omega_s t + \phi)\,, \quad \left[\omega_s^2 = \omega_0^2 - \gamma^2\right], \tag{3.15}$$

$$\gamma = \omega_0 \Rightarrow Z_{tr}(t) = (Z_1 + Z_2 t)\, e^{-\gamma t}, \tag{3.16}$$

$$\gamma > \omega_0 \Rightarrow Z_{tr}(t) = Z_1\, e^{-(\gamma - \delta)t} + Z_2\, e^{-(\gamma + \delta)t}, \quad \left[\delta = \sqrt{\gamma^2 - \omega_0^2}\right]. \tag{3.17}$$

The parameters γ and ω_0 depend on the instrument characteristics. The parameters Z_0, Z_1, Z_2 depend on the initial conditions.

The solutions $Z_{tr}(t)$ always contain a damping factor, such as $\exp(-\gamma t)$. As a consequence, for $t \to \infty$, $Z_{tr}(t) \to 0$, so that $Z(t) \to Z_{st}(t)$; the solutions of homogeneous equations have a transient behavior.

The γ parameter measures the quickness of the instrument, say the rate at which the transient solution dies away and the instrument adjusts its output to the input quantity. The parameter $\tau = 1/\gamma$ has the dimension of time and is called the *time constant* of the instrument. The damping factor is often expressed as $\exp(-t/\tau)$.

The *stationary solutions* Z_{st} of the inhomogeneous equations (3.4) and (3.5) depend on the input function $X(t)$, and describe the asymptotic behavior for $t \to \infty$.

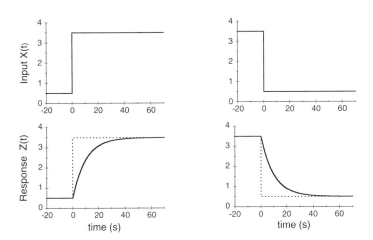

Fig. 3.10. Response of a first-order instrument to a step input. Top: two possible inputs. Bottom: the corresponding outputs (continuous lines).

Response to a Step Input

Let us consider the simple case of a step input (Fig. 3.10, top left),

$$X(t) = \begin{cases} X_0 & \text{for } t < 0, \\ X_1 & \text{for } t \geq 0, \end{cases} \qquad (3.18)$$

and suppose that, for $t < 0$, the instrument is stabilized at the stationary response $Z = (b_0/a_0)\, X_0$.

Example 3.28. A mercury thermometer is initially (for $t < 0$) in equilibrium with the environment at temperature T_0; at the time $t = 0$, it is immersed in a fluid at temperature T_1.

Example 3.29. A spring dynamometer is initially unloaded. At time $t = 0$, a mass m is suspended from the dynamometer.

Let us inquire about the behavior of the instrument for $t > 0$. After a long enough time, mathematically for $t \to \infty$, the transient solution $Z_{\text{tr}}(t)$ dies out and only the stationary solution remains,

$$Z_{\text{st}} = \frac{b_0}{a_0} X_1, \qquad (3.19)$$

which is the same for both first- and second-order instruments.

Let us now sum up the transient and stationary solutions.

For first-order instruments, the transient solution $Z_{\text{tr}}(t)$ (3.13) contains the constant Z_0, which depends on the initial conditions, and can be determined by considering the behavior of $Z(t)$ for $t = 0$. The total solution is

$$Z(t) = \frac{b_0}{a_0}(X_0 - X_1)\,\mathrm{e}^{-t/\tau} + \frac{b_0}{a_0} X_1. \qquad (3.20)$$

The output value $Z(t)$ exponentially approaches the stationary value (Fig. 3.10, bottom).

For second-order instruments, the transient solutions $Z_{\text{tr}}(t)$ (3.15) through (3.17) contain two constants (Z_0 and ϕ, or Z_1 and Z_2), which can be determined by studying the behavior of both $Z(t)$ and its derivative $\mathrm{d}Z/\mathrm{d}t$ for $t = 0$. Three different behaviors can be distinguished, depending on the relation between γ and ω_0 (Fig. 3.11).

- For $\gamma < \omega_0$, the output $Z(t)$ oscillates around the asymptotic value $Z_{\text{st}}(t)$; the more similar is γ to ω_0, the faster is the damping of oscillations.
- For $\gamma > \omega_0$, the output $Z(t)$ exponentially approaches the asymptotic value $Z_{\text{st}}(t)$, without crossing it; the more similar is γ to ω_0, the faster is the approach.
- For $\gamma = \omega_0$, one has the ideal condition of *critical damping*: the output $Z(t)$ approaches the asymptotic value $Z_{\text{st}}(t)$ in the fastest way, without crossing.

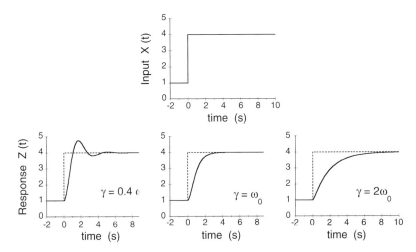

Fig. 3.11. Response of a second-order instrument to a step input. Top: input $X(t)$. Bottom: responses $Z(t)$ for $\gamma = 0.4\,\omega_0$ (left), $\gamma = \omega_0$ (center, critical damping), $\gamma = 2\,\omega_0$ (right). A static sensitivity $(b_0/a_0) = 1$ has been always assumed.

3.6 Counters

The instruments considered in previous sections, based on the direct measurement of physical quantities, such as lengths, times, and masses, do not exhaust all the requirements of a physics laboratory.

As observed in Sect. 1.5, some physical phenomena consist of events randomly distributed in time and/or space. Typical examples are radioactive decays or elementary particle collisions. However, in spite of the absolute randomness (or better, as clearer in subsequent chapters, just because of the absolute randomness), if a large enough number of events is observed, it is possible to extract regular and significant average properties. These average properties are physical quantities. Typical examples are the disintegration constants of radioactive isotopes.

Specific instruments have been devised for measuring physical quantities connected to random events. Such instruments are obviously based on the counting of random events, and are characterized by peculiar operating principles and performance parameters.

Example 3.30. The *Geiger counter* is used to count high-energy particles, such as cosmic rays or products of radioactive decays. The instrument consists of a metal vessel filled with gas and containing a tungsten filament (Fig. 3.12). Vessel and filament are at different electrical potentials. A high-energy particle can ionize one or more gas molecules, giving rise to an avalanche of secondary ionizations. The electrons so produced are collected by the filament (anode) and produce a pulse signal, which can be detected as a potential difference across the resistor R.

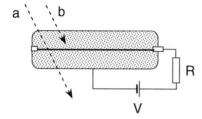

Fig. 3.12. Geiger counter. A potential difference V is maintained across vessel and filament. Tracks a and b represent the trajectories of two particles.

A counter transforms every random event into an electrical pulse. The number of electrical pulses is counted by suitable electronic circuits during a given time interval.

It is from the number of counts per unit time that one gets the values of relevant physical quantities. To this aim, because one is dealing with random events, one needs to use peculiar probabilistic methods, which are treated in Sect. 6.4. Physical quantities obtained from the counting of random events are affected by uncertainty, as is any other quantity. The uncertainty due to the randomness of the events, of purely statistical origin, is studied in Sect. 6.4 as well.

We only observe here that counters, as with any other instrument, can introduce distortions, and consequently uncertainties. Without going deep into technical details, it is worth remembering that not all the input random events are transformed into output pulses, for the following basic reasons.

(a) Some random events can completely escape detection (in the case of Geiger counters, a particle can cross the vessel without interacting with the gas molecules).
(b) Two or more events can happen simultaneously, and be counted as a single event.
(c) After the detection of a random event, a counter requires a finite amount of time (*dead time*) before being able to detect a new event; events spaced by a time interval shorter than the dead time are not distinguished.

The *detection efficiency* is the ratio between the average number of events detected in a unit time and the average number of events occurred:

$$\text{detection efficiency} = \frac{\text{counted events}}{\text{occurred events}}.$$

4 Uncertainty in Direct Measurements

Since Chap. 1, it was stated that a measurement is always affected by uncertainty. In this chapter, after a general introduction to the different causes of uncertainty in direct measurements (Sect. 4.1), the different expressions for uncertainties due to resolution (Sect. 4.2), to random fluctuations (Sect. 4.3), and to systematic errors (Sect. 4.4) are considered separately. At last, in Sect. 4.5, the different sources of uncertainty are critically compared and the rules for a standard expression of uncertainty are given.

4.1 Causes of Uncertainty

In Sect. 3.4, a first connection was made between the display resolution ΔX of an instrument and the measurement uncertainty δX. Actually, the uncertainty is in general influenced by many different factors, which cannot be reduced to the instrument characteristics:

- difficulty in defining the quantity to be measured
- operating characteristics of the instrument
- interaction between instrument and system under measurement
- interaction between instrument and experimenter
- measurement methodology
- environmental conditions

Example 4.1. The thickness of a metal foil is measured by a micrometer with display resolution $\Delta X = 0.01$ mm. The measure can be influenced by the presence of dust between the micrometer rods and the foil (*interaction between instrument and system*). If the micrometer has been calibrated at 20°C and is used to perform measurements at much lower or much higher temperatures, nonnegligible errors can be introduced (*environment influence*). The thickness of the metal foil can be different in different parts of the foil (*difficulty in defining the physical quantity*).

Example 4.2. The period of a pendulum is measured by a manual stopwatch with display resolution $\Delta t = 0.01$ s. The measure will depend on the quickness of reflexes of the experimenter (*interaction between instrument and experimenter*). Moreover, the result can be different according to whether the duration of a single period is measured, or the duration of ten periods is measured and then divided by ten (*measurement methodology*).

The different factors that can influence a measure can be classified within three main categories:

- measurement resolution
- random fluctuations (or "random errors")
- systematic errors

These three causes of uncertainty are analyzed separately in the following three sections. A unified procedure for comparing and combining the uncertainties due to different factors is introduced in Sect. 4.5.

It is worth noting that the terms *random error* and *systematic error* have here a particular meaning, and should not be confused with trivial mistakes (*parasitic errors*) that are due to carelessness or inexperience, such as exchanging a mass of 50 g for a mass of 100 g on a two-pan balance, the wrong reading of a display, or a calculus error in an indirect measurement. Although not at all negligible, parasitic errors are not susceptible to formal treatment, and are not taken into consideration in the following. Their presence has to be avoided by carefully planning and performing the experiments.

4.2 Measurement Resolution

A first cause of uncertainty is the measurement resolution.

Instrument Resolution and Measurement Resolution

In Sect. 3.4, the display resolution of an instrument was introduced; let us label it here ΔX_{inst}. It is convenient to introduce the more general concept of measurement resolution ΔX_{meas}. In many cases, the measurement resolution corresponds to the display resolution of the instrument.

Example 4.3. The period of a pendulum is measured by a stopwatch with display resolution $\Delta t_{inst} = 0.01$ s. The measurement resolution is equal to the instrument display resolution, $\Delta t_{inst} = \Delta t_{meas} = 0.01$ s.

Sometimes, however, the measurement resolution can be reduced with respect to the instrument resolution by suitable methodologies. Typically, instead of measuring the quantity X, one measures a multiple nX. The instrument display resolution ΔX_{inst} refers now to the value nX; the measurement resolution of X is n times smaller than the instrument resolution: $\Delta X_{meas} = \Delta X_{inst}/n$.

Example 4.4. The period $\mathcal{T}$ of a pendulum is again measured by a stopwatch with display resolution $\Delta t_{inst} = 0.01$ s. The duration ΔT of $n = 10$ consecutive oscillations is measured, and the period is calculated as $\mathcal{T} = \Delta T/10$. The measurement resolution of the period $\mathcal{T}$ is thus $n = 10$ times smaller than the instrument resolution: $\Delta t_{meas} = \Delta t_{inst}/10 = 0.001$ s.

Uncertainty Due to Resolution

The result of a measurement is an interval of values whose extension corresponds to the measurement resolution. If $X_{\min}$ and $X_{\max}$ are the lower and upper limits of the interval, respectively:

$$X_{\min} \leq X \leq X_{\max} . \qquad (4.1)$$

All X values within the resolution interval are equivalent. One can also say that they are distributed according to a uniform probability distribution. The graphical representation of the uniform probability distribution is a rectangle of width ΔX and unit area (Fig. 4.1). The concepts of probability and probability distributions, here introduced in a heuristic way, are systematically treated in Chaps. 5 and 6.

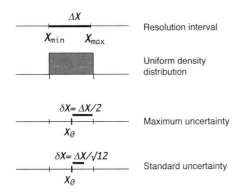

Fig. 4.1. Graphical representation of the link between resolution interval and uniform probability distribution, as well as of the difference between maximum uncertainty and standard uncertainty.

As anticipated in Sect. 3.4, the result of a measurement is generally expressed as

$$X = X_0 \pm \delta X , \qquad (4.2)$$

where:
- X_0 is the central value of the resolution interval
- δX is the uncertainty.

A reasonable choice for the uncertainty δX is the half-width of the resolution interval (Fig. 4.1):

$$\delta X_{\max} = \Delta X / 2 . \qquad (4.3)$$

The uncertainty defined in (4.3) includes all possible values of X within the resolution interval; that's why it is called *maximum resolution uncertainty* $\delta X_{\max}$.

In Sect. 4.5, it is shown that, in order to guarantee a consistent comparison with uncertainty due to random errors, a conventional choice different from (4.3) is preferable, the so-called *standard resolution uncertainty*

$$\delta X_\text{res} = \Delta X / \sqrt{12}\,. \tag{4.4}$$

Maximum uncertainty and standard uncertainty have different meanings, and their numerical values are different. When quoting a value of uncertainty due to resolution, it is thus necessary to specify whether the value δX refers to maximum or standard uncertainty.

Example 4.5. The period T of a pendulum is measured by a stopwatch with resolution $\Delta t = 0.01$ s. The display shows the value 1.75 s. The measure is expressed as $T = (1.75 \pm 0.005)$ s if the maximum uncertainty is quoted, according to (4.3), or as $T = (1.75 \pm 0.003)$ s if the standard uncertainty is quoted, according to (4.4).

One can easily verify that, in the above example, the uncertainty calculated through (4.4) has been expressed as a rounded value: $0.01/\sqrt{12} = 0.00288675 \simeq 0.003$. A measurement uncertainty should always be expressed with one or two significant digits. A larger number of significant digits is meaningless. A thorough treatment of significant digits and rounding procedures is given in Appendix A.1.

4.3 Random Fluctuations

A second cause of uncertainty is due to random fluctuations (which for historical reasons are often called "random errors").

Repeated Measurements

Let us consider a physical quantity constant in time, and repeat its measurement many times. Two cases can occur:

(a) All measures fall within the same resolution interval (Fig. 4.2 *a*).
(b) Different measures randomly fall within different resolution intervals (Fig. 4.2 *b*).

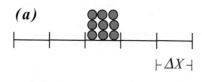

Fig. 4.2. Schematic representation of two different situations that can be encountered when repeating the measurement of a physical quantity (ΔX is the measurement resolution).

Example 4.6. The height of a paper sheet is measured by a ruler with resolution $\Delta X = 1$ mm. When the measurement is repeated, all values fall within the same resolution interval (case (a)).

Example 4.7. The period of a pendulum is measured by a manual stopwatch with resolution $\Delta t = 0.01$ s. When the measurement is repeated, different values are randomly obtained, falling within different resolution intervals (case (b)).

In case (a), the uncertainty δX only depends on the resolution ΔX, and is quoted as $\delta X_{\max} = \Delta X/2$ (maximum uncertainty) or $\delta X_{\text{res}} = \Delta X/\sqrt{12}$ (standard uncertainty).

In case (b), the discrepancies between different measures of the same quantity depend on the simultaneous and random influence of many small factors, each one acting independently of the others on the single measure. For example:

– Reading errors: inadequacy of the eye separating power, parallax errors, interpolation errors, synchronization errors, and so on
  Background noise, say the effect of very small variations of influence quantities, like temperature, pressure, humidity, vibrations, and so on
– Inversion errors (difference between results of measurements performed while the value of the measured quantity is increasing or decreasing)

In case (b), the measurements are said to be affected by random fluctuations (or random errors), which cause an uncertainty larger than the uncertainty due to resolution.

Actually, one can assume that random fluctuations are always present. Their effect is, however, unnoticed when it is globally smaller than the measurement resolution (Fig. 4.3, a). If, however, the resolution is suitably reduced, the effect of random fluctuations becomes dominant (Fig. 4.3, b).

(a)

(b)

Fig. 4.3. Random fluctuations and different measurement resolutions.

In the remainder of this section, the effect of random fluctuations is considered as larger than the effect of resolution, and the following question is faced: how can the uncertainty due to random fluctuations be quantitatively evaluated in order to express the measure as $X_0 \pm \delta X$?

To answer this question, it is convenient to introduce some techniques for representing and treating experimental data, such as histograms and their statistical parameters, as well as the concept of limiting distribution.

Histograms

Let us suppose that a constant physical quantity X has been measured N times. Each measure is labeled by an index i ($i = 1, \ldots, N$). Different measures x_i can fall within the same resolution interval. Histograms are a convenient way to describe this situation (Fig. 4.4 and Appendix A.4). Every column of the histogram has a width ΔX equal to the measurement resolution. The columns are labeled by an index j. The height of the jth column is proportional to the number n_j^* of values within the jth interval (Fig. 4.4, left). If $\mathcal{N}$ is the number of columns, it is easy to check that

$$\sum_{j=1}^{\mathcal{N}} n_j^* = N \ . \tag{4.5}$$

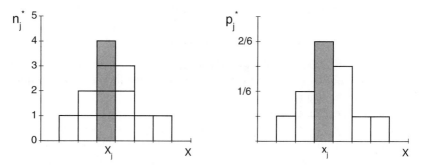

Fig. 4.4. Left: histogram for $N = 12$ measures; the height of the jth column gives the number n_j^* of measures having value x_j. Right: the same histogram, normalized in height; the height of the jth column gives the sample frequency $p_j^* = n_j^*/N$.

The heights of the histogram columns depend on the total number N of measures. To compare histograms based on different numbers N of measures, it is convenient to introduce height normalization. In a *height-normalized histogram* (Fig. 4.4, right), the height of each column is proportional to the corresponding *sample frequency*

$$p_j^* = n_j^*/N \ . \tag{4.6}$$

One can easily verify that, for whichever number N of measures,

$$\sum_{j=1}^{\mathcal{N}} p_j^* = 1 \ . \tag{4.7}$$

The meaning of the term *sample* is clarified below in this section. In this book, the asterisk (*) conventionally labels sample quantities.

Sometimes it is necessary to compare two sets of values of the same quantity obtained from measurements performed with different resolutions, so that the corresponding histograms have columns of different widths ΔX. In these cases, it is convenient to introduce the area normalization. In an *area-normalized histogram* (Fig. 4.5), the height of each column is proportional to the *sample density*

$$f_j^* = \frac{n_j^*}{N \, \Delta X_j} \, . \tag{4.8}$$

It is easy to verify that, for whichever number N of measures, the total area of the columns is one:

$$\sum_{j=1}^{\mathcal{N}} f_j^* \, \Delta X_j = 1 \, . \tag{4.9}$$

The sample frequency p^* is always dimensionless, while the sample density f^* defined in (4.8) always has the dimension of $1/X$.

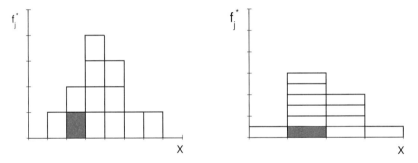

Fig. 4.5. Two area-normalized histograms for the same set of measures. In the right-handside histogram, data have been grouped into columns of width double than in the left-handside histogram.

Statistical Parameters of Histograms

A histogram contains all the amount of information that can be obtained by measuring a physical quantity N times. Sometimes, however, it is sufficient to synthesize the main properties of a histogram by a few numerical parameters. Only two such parameters are considered here, the first one giving the average position of the histogram on the x-axis, and the second one giving the dispersion of the measured values. These two parameters are sufficient for conveniently expressing the results of measurements affected by random fluctuations.

Position Parameter

The position of the histogram (Fig. 4.6, left) is given by the *sample mean* m^*, corresponding to the arithmetic average of the N values x_i:

$$m^* = \frac{1}{N} \sum_{i=1}^{N} x_i \,. \tag{4.10}$$

Alternatively to (4.10), one can sum over the $\mathcal{N}$ histogram columns, characterized by the sample frequencies p_j^*:

$$m^* = \frac{1}{N} \sum_{j=1}^{\mathcal{N}} x_j n_j^* = \sum_{j=1}^{\mathcal{N}} x_j \frac{n_j^*}{N} = \sum_{j=1}^{\mathcal{N}} x_j p_j^* \,. \tag{4.11}$$

Different equivalent symbols are chosen to represent mean values,

$$m^* \equiv m_x^* \equiv \langle x \rangle^* \equiv \mathbf{m}^*[x] \,,$$

in order to guarantee the better readability of the current formula.

Dispersion Parameters

To measure the dispersion of a histogram, one first defines the *deviation* s_i of each value x_i from the sample mean:

$$s_i = x_i - m^* \,. \tag{4.12}$$

The average value of the deviations s_i is unsuited to measure the dispersion, because it is identically zero:

$$\langle s \rangle = \frac{1}{N} \sum_{i=1}^{N} s_i = \frac{1}{N} \sum_{i=1}^{N} (x_i - m^*) = \frac{1}{N} \sum_{i=1}^{N} x_i - \frac{1}{N} N m^* = 0 \,. \tag{4.13}$$

A conventional measure of the histogram dispersion is the *sample variance* D^*, defined as the average value of the squared deviations s_i^2. As with the sample mean, the sample variance can also be calculated by summing over all N measured values or over the $\mathcal{N}$ histogram columns:

$$D^* = \langle (x_i - m^*)^2 \rangle = \frac{1}{N} \sum_{i=1}^{N} (x_i - m^*)^2 = \sum_{j=1}^{\mathcal{N}} (x_j - m^*)^2 p_j^* \,. \tag{4.14}$$

The sample variance has the dimension of the square of the quantity X. The square root of the sample variance, the *sample standard deviation* σ^*, has the same dimension of X, and can be directly visualized on the graph (Fig. 4.6, right):

$$\sigma^* = \sqrt{D^*} = \sqrt{\frac{1}{N} \sum_{i=1}^{N} (x_i - m^*)^2} = \sqrt{\sum_{j=1}^{\mathcal{N}} (x_j - m^*)^2 p_j^*} \,. \tag{4.15}$$

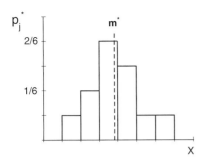 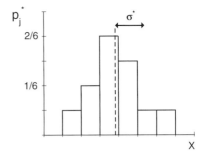

Fig. 4.6. Sample mean m^* (left) and sample standard deviation σ^* (right) of a histogram.

Example 4.8. The period of a pendulum is measured $N = 20$ times by a stopwatch with resolution $\Delta t = 0.01\,\text{s}$. The results, grouped in histogram columns, are listed in Table 4.1. Sample mean, sample variance, and sample standard deviation are, respectively, $m^* = 1.25\,\text{s}$, $D^* = 2.25 \times 10^{-4}\,\text{s}^2$, and $\sigma^* = 1.5 \times 10^{-2}\,\text{s}$.

Table 4.1. Distribution of the values of the period of a pendulum (Example 4.8).

	T_j [s]	n_j^*	p_j^*	f_j^* [s^{-1}]
	1.22	1	0.05	5
	1.23	2	0.1	10
	1.24	4	0.2	20
	1.25	6	0.3	30
	1.26	3	0.15	15
	1.27	3	0.15	15
	1.28	1	0.05	5
$\sum$		20	1.00	100

One can easily verify that the variance D^* is the difference between the average of the squared values x_i^2 and the square of the average value $\langle x \rangle$:

$$D^* = \frac{1}{N} \sum_{i=1}^{N} (x_i - \langle x \rangle)^2 = \frac{1}{N} \sum_{i=1}^{N} [x_i^2 - 2x_i \langle x \rangle + \langle x \rangle^2]$$

$$= \frac{1}{N} \sum_{i=1}^{N} x_i^2 - \langle x \rangle^2 = \langle x^2 \rangle - \langle x \rangle^2 \,. \tag{4.16}$$

The expression (4.16), alternative to (4.14), is sometimes useful to speed up the calculations of variance D^* and standard deviation σ^*.

Limiting Histogram and Limiting Distribution

It has just been shown that the results of N measurements affected by random fluctuations can be represented by a histogram, or, in a more synthetic although less complete way, by two parameters: sample mean m^* and sample standard deviation σ^*.

Let us now suppose that a new set of N measurements is performed on the same quantity X; one expects to obtain a different histogram, with different values m^* and σ^*. By repeating other sets of N measurements, one again obtains different histograms and different values m^* and σ^*. The histogram relative to N measurements and its statistical parameters m^* and σ^* have thus a random character.

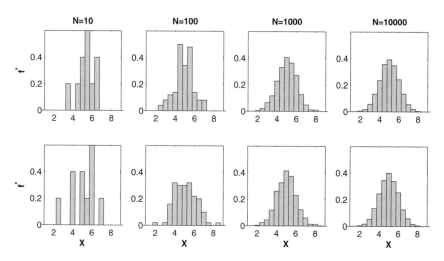

Fig. 4.7. Eight area-normalized histograms, relative to different measurements of the same quantity. The two left histograms ($N = 10$ measurements) are very different. When the number of measurements increases from $N = 10$ to $N = 10000$ (from left to right), the histograms progressively lose their random character and tend to assume a well-defined shape.

It is, however, a matter of experience that when the number N of measurements increases, the histograms tend to assume a similar shape (Fig. 4.7); correspondingly, the differences between the values m^* and σ^* of different histograms tend to reduce. This observed trend leads to the concept of *limiting histogram*, towards which the experimental histograms are supposed to tend when the number of measurements N increases, ideally for $N \to \infty$.

The limiting histogram is clearly an abstract idea, whose existence cannot be experimentally verified (N is necessarily finite). Assuming the existence of a limiting histogram corresponds to postulating the existence of a regularity in natural phenomena, which justifies the enunciation of physical laws of general validity on the grounds of a limited number of experimental observations.

In many cases, although not always, the histograms of measurements affected by random fluctuations tend to a symmetric "bell" shape when N increases (Fig. 4.7, right). The limiting histogram is then assumed to have a bell shape. It is convenient to describe the bell shape of the limiting histogram by a mathematical model, expressed in terms of a continuous function. To this aim, the further approximation of shrinking to zero the width of the histogram columns is made: $\Delta x \to 0$. By that procedure, the limiting histogram is substituted by a *limiting distribution*, corresponding to a continuous function of continuous variable $f(X)$.

The Normal Distribution

According to both experimental observations and theoretical considerations, the bell-shaped behavior of the limiting distribution is best represented by the *normal distribution*, also called *Gaussian distribution*, after the name of the German mathematician C. F. Gauss (1777–1855):

$$f(x) = \frac{1}{\sigma\sqrt{2\pi}} \exp\left[-\frac{(x-m)^2}{2\sigma^2}\right]. \tag{4.17}$$

The parameters m and σ in (4.17) have the same dimension of the variable x. It easy to verify that m gives the position of the distribution on the x-axis, whereas σ depends on the width of the distribution (Fig. 4.8).

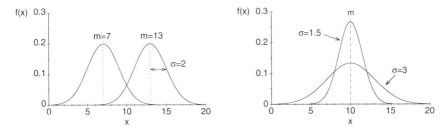

Fig. 4.8. Normal distribution (4.17). Left: two distributions with the same standard deviation σ and different means m. Right: two distributions with the same mean m and different standard deviations σ.

The function $f(x)$ in (4.17) is dimensionally homogeneous to the sample density f_j^* defined in (4.8). The normal distribution is the limiting distribution of an area-normalized histogram (Fig. 4.7) for both $N \to \infty$ (number of measurements) and $\mathcal{N} \to \infty$ (number of columns, corresponding to $\Delta x_j \to 0$). One can show that m and σ are the asymptotic values, for $N \to \infty$, of sample mean m^* and sample standard deviation σ^*, respectively. To this aim, one substitutes $p^* = f^* \Delta x$ in the expressions (4.11) and (4.14) of m^* and D^*, respectively. The limit for $\mathcal{N} \to \infty$ is obtained by substituting sums with integrals:

$$m^* = \sum_{j=1}^{\mathcal{N}} x_j f_j^* \Delta x_j \quad \to \quad m = \int_{-\infty}^{+\infty} x f(x) \, \mathrm{d}x , \tag{4.18}$$

$$D^* = \sum_{j=1}^{\mathcal{N}} (x_j - m^*)^2 f_j^* \Delta x_j \quad \to \quad D = \int_{-\infty}^{+\infty} (x - m)^2 f(x) \, \mathrm{d}x . \tag{4.19}$$

The square root of the variance D is the standard deviation $\sigma = \sqrt{D}$. The parameters m and σ are thus the average value and the standard deviation, respectively, of the limiting distribution.

By substituting the sum (4.9) with an integral, one can easily show that, like the area of area-normalized histograms, the integral of normal distributions is also one:

$$\sum_{j=1}^{\mathcal{N}} f_j^* \Delta X_j = 1 \quad \to \quad \int_{-\infty}^{+\infty} f(x) \, \mathrm{d}x = 1 . \tag{4.20}$$

The properties of the normal distribution are thoroughly analyzed in Chap. 6. There it is demonstrated that, for any values of m and σ (Fig. 4.9):

The area between $m - \sigma$ and $m + \sigma$ is 0.683
The area between $m - 2\sigma$ and $m + 2\sigma$ is 0.954
The area between $m - 3\sigma$ and $m + 3\sigma$ is 0.997
The total area is 1

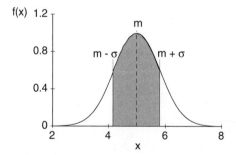

Fig. 4.9. Normal distribution with $m = 5$ and $\sigma = 0.8$. The area under the curve between $x = m - \sigma$ and $x = m + \sigma$ is always 0.68, for any values m and σ.

As with the limiting histogram, the limiting distribution is also an abstract idea, defined for $N \to \infty$ and $\Delta x \to 0$. The limiting distribution cannot be completely determined from experiment. One can only obtain an approximate knowledge, based on the measurements actually performed. The higher is the number N of measurements, the better is the approximation.

The normal distribution (4.17) is actually the best limiting distribution in many cases, but can never be considered an exact model, in particular in the tail regions. In fact, the normal distribution is defined over all the real axis ($-\infty < x < +\infty$), and many physical quantities, such as the period of a pendulum, can only assume positive values; in addition, the normal distribution is never zero for any value of x.

In spite of these limitations, the hypothesis of existence of a limiting distribution represents a powerful tool for quantitatively evaluating the results of a finite set of repeated measurements of a physical quantity. On more general grounds, the hypothesis that a regular universal behavior can be extrapolated from a necessarily finite number of observations can be considered at the very basis of the scientific method.

Estimating the Parameters of the Normal Distribution

The results of N repeated measurements of a physical quantity can be considered as a finite sample of a limiting distribution that ideally corresponds to an infinite number of measurements.

It is impossible to exactly determine the parameters m and D of a limiting distribution from a finite sampling. Actually, the sample parameters m^* and D^* have a random character, because their values depend on the particular sample. One can instead reasonably estimate the parameters m and D of the limiting distribution from the parameters m^* and D^* of a given sample of N measures. The estimation of parameters is thoroughly treated in Sect. 7.3. Some relevant results are anticipated here.

Let us suppose that N values of a physical quantity have been measured.

(a) The best estimate $\tilde{m}$ of the mean m of the limiting distribution is the sample mean m^*:

$$\tilde{m} = m^* = \frac{1}{N} \sum_{i=1}^{N} x_i . \qquad (4.21)$$

(b) The best estimate $\tilde{D}$ of the variance D of the limiting distribution is not the sample variance D^*, but

$$\tilde{D} = \frac{N}{N-1} D^* = \frac{1}{N-1} \sum_{i=1}^{N} (x_i - m^*)^2 . \qquad (4.22)$$

It is evident from (4.22) that $\tilde{D} > D^*$, the difference decreasing when N increases. The sample variance D^* underestimates the limiting variance D, because it considers the deviations from the sample mean m^* instead of the deviations from the limiting mean m.

Starting from (4.22), one can also evaluate the best estimate $\tilde{\sigma}$ of the limiting standard deviation $\sigma = \sqrt{D}$:

$$\tilde{\sigma} = \sqrt{\frac{N}{N-1}} \sigma^* = \sqrt{\frac{1}{N-1} \sum_{i=1}^{N} (x_i - m^*)^2} \,. \tag{4.23}$$

The value $\tilde{\sigma}$ of (4.23) is often called the experimental standard deviation, to distinguish it from the sample standard deviation σ^*.

Example 4.9. To better grasp the difference between σ^* and $\tilde{\sigma}$, let us consider three sets containing $N = 6$, 4, and 2 values x_i, respectively, with the same sample mean $m^* = 5$ and standard deviation $\sigma^* = 1$. One can easily verify that the experimental standard deviation $\tilde{\sigma}$ (4.23) increases when N decreases, and becomes progressively larger than $\sigma^* = 1$.

N	x_i	m^*	σ^*	$\tilde{\sigma}^*$
6	4,4,4,6,6,6	5	1	1.095
4	4,4,6,6	5	1	1.155
2	4,6	5	1	1.410

Example 4.10. If only one measurement is performed, say $N = 1$, only one value x_1 is obtained, so that $m^* = x_1$ and $\sigma^* = 0$. According to (4.23), the experimental standard deviation $\tilde{\sigma}$ cannot be determined. There is no possibility of estimating the width of the limiting distribution from only one measurement.

Distribution of Sample Means

The limiting distribution considered above describes the dispersion of the single values of a quantity X due to random fluctuations.

Let us now consider M different samples, each one consisting of N measures. The sample means m^* of each one of the M samples

$$m_1^*, \quad m_2^*, \quad m_3^*, \quad \ldots \quad m_M^*,$$

can be represented in a histogram, the histogram of the sample means.

When the number M of samples increases, the histogram of sample means tends to an increasingly regular shape. One is naturally led to the concept of a limiting histogram of the sample means (for $M \to \infty$), and of a limiting distribution $g(m^*)$ of the sample means.

One can reasonably expect that the limiting distribution $g(m^*)$ of the sample means is narrower than the limiting distribution $f(x)$ of the single measures. One can also expect that the larger the number N of measures on which each sample mean is calculated, the narrower is the distribution of sample means (Fig. 4.10).

It is demonstrated in Sect. 6.6 that:

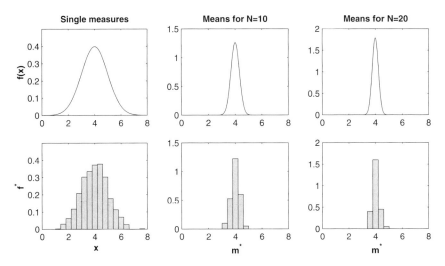

Fig. 4.10. Top left: limiting normal distribution of single measures, with $m = 4$ and $\sigma = 1$ (arbitrary units). Top (center and right): corresponding limiting distributions of sample means m^*, from $N = 10$ and $N = 20$ measures, respectively. Bottom left: an experimental sample of $N = 1000$ single measures x. Bottom (center and right): samples of $M = 100$ mean values m^* from $N = 10$ measures and $M = 50$ mean values m^* from $N = 20$ measures, respectively.

(a) For any limiting distribution of single values $f(x)$ (not necessarily normal), the distribution $g(m^*)$ of the sample means is to a good approximation normal, if the number N of measures, over which each mean m^* is calculated, is large enough.

It is further demonstrated in Sect. 7.2 that:

(b) The mean $\mathbf{m}[m^*]$ of the limiting distribution of sample means is equal to the mean m of the limiting distribution of single measures:

$$\mathbf{m}[m^*] = m \, . \tag{4.24}$$

(c) The variance $\mathbf{D}[m^*]$ and the standard deviation $\sigma[m^*]$ of the limiting distribution of sample means are connected to the variance D and the standard deviation σ of the limiting distribution of single measures, respectively, by the relations:

$$\mathbf{D}[m^*] = \frac{1}{N} D, \quad \sigma[m^*] = \frac{1}{\sqrt{N}} \sigma \, , \tag{4.25}$$

where N is the number of measurements of each sample.

Uncertainty Due to Random Fluctuations

One can now finally solve the problem of expressing in the standard form $X_0 \pm \delta X$ the result of N repeated measurements affected by random fluctuations.

If the limiting distribution $f(x)$ of single measures were perfectly known, one could quote as the *true value* of the physical quantity the mean of the limiting distribution, $X_0 = m$. The uncertainty due to random fluctuations would then be zero, $\delta X_{\text{cas}} = 0$ (the index "cas" means "casual", synonymous with "random").

Actually, the limiting distribution is never known. One can however assume its existence, and consider the N measures as a limited sample. Because the sample mean m^* is the best estimate of the mean m of the limiting distribution, it is reasonable to assume the sample mean m^* as the central value X_0:

$$X_0 = m^* . \tag{4.26}$$

The sample mean m^* is a random variable, whose value would randomly change if the sampling were repeated. The uncertainty δX_{cas} depends on the randomness of the sample mean m^*, say on the width of the distribution of sample means.

It has been shown in Sect. 4.2 that the uncertainty due to resolution can be described by a rectangular distribution of width ΔX (Fig. 4.1), so that a maximum uncertainty could be defined, $\delta X_{\max} = \Delta X/2$.

The limiting distribution of sample means is instead normal, without lower and upper limits, so that a maximum uncertainty cannot be defined. The width of the distribution of sample means can be measured by the standard deviation $\sigma[m^*]$, so that, for random fluctuations, one can only define a *standard uncertainty*, proportional to the standard deviation:

$$\delta X_{\text{cas}} \propto \sigma[m^*] . \tag{4.27}$$

By convention, the standard uncertainty δX_{cas} due to random fluctuations is assumed equal to the standard deviation of the distribution of sample means:

$$\delta X_{\text{cas}} = \sigma[m^*] . \tag{4.28}$$

Experimental Evaluation of Uncertainty

The procedure for evaluating the uncertainty due to random fluctuations from a finite set of measures is based on the following logical sequence.

(a) The uncertainty due to random fluctuations is measured by the standard deviation of the distribution of sample means: $\delta X_{\text{cas}} = \sigma[m^*]$.
(b) The standard deviation of the distribution of sample means $\sigma[m^*]$ is in turn connected to the standard deviation σ of the limiting distribution of single values by the relation (4.25), $\sigma[m^*] = \sigma/\sqrt{N}$.
(c) The standard deviation σ of the limiting distribution of single values can be estimated, from the finite sample, through (4.23).

This logical sequence leads to

$$\delta X_{\text{cas}} = \tilde{\sigma}[m^*] = \frac{1}{\sqrt{N}} \sigma = \frac{1}{\sqrt{N}} \sqrt{\frac{N}{N-1}} \sigma^*$$

$$= \sqrt{\frac{1}{N(N-1)} \sum_{i=1}^{N} (x_i - m^*)^2} \ . \tag{4.29}$$

The estimate $\tilde{\sigma}[m^*]$ is called the *experimental standard deviation of the mean*. It is evident that at least two measurements ($N > 1$) are necessary to evaluate δX_{cas}.

Let us now go deeper into the meaning of (4.28).

It has been stated that the area under the normal distribution, included between the values $x = m - \sigma$ and $x = m + \sigma$, is 0.68. If the normal distribution of sample means, centered in m and with standard deviation $\sigma[m^*]$, were a priori known, one would know that the mean m^* of any sample has a probability 0.68, say 68%, of falling within the interval $m^* \pm \sigma[m^*]$.

In real cases, however, it is the sample values m^* e σ^* that are known, and one estimates m. The problem is thus to calculate the probability that m is within the interval $m^* \pm \delta X_{\text{cas}}$, where δX_{cas} is estimated through (4.29), and not exactly known.

This problem is not trivial, and is solved in Sect. 9.2. Only the main results are anticipated here. The probability that m is within the interval $m^* \pm \delta X$ is 68% only for $N \to \infty$. If N is large, the probability can be assumed, to a good accuracy, to be 68%. If, however, N is small, the probability can be significantly less than 68%. It is thus good practice, when expressing the uncertainty due to random fluctuations, to always quote the number N of measurements performed.

Note. In this Sect. 4.3, the following convention has been adopted: the parameters concerning limiting distributions and samples are m, D, σ and m^*, D^*, σ^*, respectively, whereas $\tilde{m}, \tilde{D}, \tilde{\sigma}$ are the corresponding estimates.

4.4 Systematic Errors

Systematic errors, say errors that modify the measure by the same amount whenever the measurement is repeated, have been introduced in Sect. 3.4 when considering the performance of instruments. Systematic errors can, however, be due not only to the instrument, but also to the measurement procedures (Sect. 4.1).

Example 4.11. The period of a pendulum is measured by a retarding stopwatch. When the measurements are repeated, all measures are reduced by the same amount (systematic error due to the wrong calibration of the instrument).

Example 4.12. The period of a pendulum is measured by a manual stopwatch, The measure depends on the difference between the reaction time of the experimenter when the stopwatch is started and stopped. The difference of reaction times can have both a random and a systematic component, the former giving rise to the dispersion of measures, and the latter giving rise to a constant variation of the measures.

The large variety of situations giving rise to systematic errors prevents an exhaustive formal treatment. The search for systematic errors and their elimination is a particularly difficult and delicate task. In the following, some relevant cases are considered and commented on, without any claim of generality.

Measurements Repeated in Fixed Conditions

Let us suppose that the measurement of a physical quantity is repeated many times in the same conditions, say by the same experimenter, by the same procedure and instrument, at the same site, and within a limited time interval. Each measure is affected by both random fluctuations and systematic errors. As pointed out in Sect. 4.3, random fluctuations cannot be completely eliminated, but it is always possible to quantitatively evaluate their effect by repeating the measurement, and express it through an uncertainty δX_{cas}. On the contrary, systematic errors cannot be evidenced by repeating the measurement in the same experimental conditions. They can be singled out only by means of an accurate analysis of the instrument performance and of the measurement methodology. Let us consider some examples.

1. Sometimes, a careful analysis leads us to find a systematic error of the measurement methodology. The error can then be eliminated by modifying the methodology, or compensated by suitably correcting the measure. The correction can always be affected by an uncertainty δX_{sys}, which contributes to the global uncertainty of the measure.

Example 4.13. One wants to measure the acceleration of gravity g by exploiting the relation $T = 2\pi(\ell/g)^{1/2}$ between period T and length ℓ of a pendulum, in the small oscillation approximation (Experiment E.5 in Appendix E). The pendulum is a metal cylinder suspended by a string. Length ℓ and period T are directly measured, and then $g = (2\pi/T)^2 \ell$ is calculated. A careful analysis shows, however, that the period depends also on the oscillation amplitude and on the distribution of the cylinder mass. The relation $T = 2\pi(\ell/g)^{1/2}$ is approximate, and the value g is thus affected by a systematic error. A more exact, although still approximate, relation is $T = 2\pi \left(I/mg\ell \right)^{1/2} [1 + (1/4) \sin^2(\theta_0/2)]$, where I is the moment of inertia of the cylinder with respect to the oscillation axis and θ_0 is the oscillation amplitude.

2. In some situations, the leading causes of systematic errors can be singled out, but it is impossible to exactly evaluate and correct their effect. In that case, one tries to estimate the extent of the possible systematic errors and express it as an uncertainty δX_{sys}.

Example 4.14. One wants to measure the electric current I in a branch of an electric circuit. To this aim, an ammeter is inserted in series in the circuit branch. The instrument has a nonzero internal resistance R_i and modifies the characteristics of the circuit. The measured current is $I_m < I$, the difference $I_m - I$ being a systematic error. In principle, if the internal resistance R_i and the circuit characteristics are known, the systematic error can be evaluated and corrected. In practice, if the circuit characteristics are not completely known, it is always possible to estimate an upper limit of the systematic error.

3. Sometimes, the systematic error is due to an insufficient definition of the quantity to be measured.

Example 4.15. One wants to measure the dependence of a pendulum period T on the oscillation amplitude θ_0 (Experiment E.6 in Appendix E). To this aim, pairs of values (θ_0, T) are measured for different values θ_0. However, the amplitude θ_0 progressively reduces in time, mainly due to air resistance. For large θ_0 values, the reduction can be nonnegligible within one period T. The amplitude of oscillation is in this case not well defined. We can face this situation by considering the actual measure of the amplitude as affected by a systematic error, and attribute to its value a suitable uncertainty. The uncertainty could be estimated by measuring the reduction of θ_0 within an oscillation.

Measurements Repeated in Different Conditions: Discrepancy

The search for systematic errors, or at least an estimate of their extent, is facilitated by comparing measurements carried on in different conditions (different experimenters, procedures, instruments, sites, and so on). For the sake of simplicity, the comparison of only two measurements performed in different conditions, labeled A and B, is considered at first.

Example 4.16. Two experimenters, A and B, measure the period of a pendulum by a manual stopwatch (Experiment E.2 in Appendix E). The systematic component of reaction times can be different for the two experimenters.

Example 4.17. The elastic constant k of a spring is measured by two procedures. Procedure A consists of measuring the ratio between applied force F and deformation x, exploiting the relation $F = kx$ (Experiment E.3 in Appendix E). Procedure B consists of attaching a mass to an end of the spring, and measuring the relation between mass m and oscillation period T, exploiting the relation $T = 2\pi(m/k)^{1/2}$ (Experiment E.4 in Appendix E). Both procedures refer to an *indirect* measurement of the elastic constant k; the evaluation of uncertainty in indirect measurements is considered in Chap. 8.

The results of measurements performed by two different procedures A and B can be expressed as

$$X_A \pm \delta X_A , \quad X_B \pm \delta X_B , \tag{4.30}$$

where the uncertainties δX_A and δX_B depend, according to the different situations, on resolution (Sect. 4.2) or on random fluctuations (Sect. 4.3), and can include contributions due to already found systematic errors.

Sometimes one of the two uncertainties is much larger than the other one, $\delta X_A \gg \delta X_B$; this is a typical situation in didactic laboratories, where X_A is measured by students with rudimentary instrumentation whereas X_B is a reference value, quoted in specialized journals or books. In this case, one can set $\delta X_B = 0$, and the following considerations are simplified.

The *discrepancy* of the two measures is the modulus of their difference: $|X_A - X_B|$. One could suppose that the discrepancy is always due to systematic errors in at least one of the two measurements. Actually, before drawing definitive conclusions, one has to compare the discrepancy with the uncertainties of the two measures. An exhaustive treatment of this topic is practically impossible in view of the large amount of different possible situations. Let us only give here some simple limiting examples.

1. Both uncertainties of X_A and X_B are due to resolution, and described by rectangular distributions of widths ΔX_A and ΔX_B, respectively. If the discrepancy is comparable or smaller than the sum of the maximum uncertainties, say to the half-sum of resolutions (Fig. 4.11, upper left),

$$|X_A - X_B| \leq \frac{\Delta X_A + \Delta X_B}{2} , \tag{4.31}$$

   then the values X_A and X_B can be considered consistent. The discrepancy $|X_B - X_A|$ can be attributed to the uncertainty of the single measures and cannot help in evaluating the systematic errors.

2. Both uncertainties of X_A and X_B are still due to resolution, but now the discrepancy is larger than the sum of the maximum uncertainties, say of the half-sum of resolutions (Fig. 4.11, upper right),

$$|X_A - X_B| > \frac{\Delta X_A + \Delta X_B}{2} . \tag{4.32}$$

   The values X_A and X_B are now inconsistent, and the discrepancy can generally be attributed to the effect of systematic errors in at least one of the two measurement procedures.

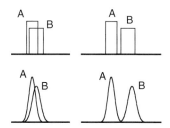

Fig. 4.11. Schematic comparison of the measures of a quantity X obtained by two different procedures, A and B. The uncertainties are due to resolution in the upper panels (rectangular distributions) and to random fluctuations in the lower panels (normal distributions). Left and right panels refer to consistency and inconsistency cases, respectively.

3. Both uncertainties of X_A and X_B are due to random fluctuations, and are described by normal distributions (4.17). There is no maximum uncertainty in this case. The probability of finding values outside the interval $\pm 3\sigma$ centered on the mean m is, however, negligible. As a consequence, if the discrepancy is larger than the sum of the two "3σ" intervals (Fig. 4.11, lower right),

$$|X_A - X_B| > 3\sigma_A + 3\sigma_B , \qquad (4.33)$$

we can consider the values X_A and X_B as inconsistent, and attribute the discrepancy to the effect of systematic errors in at least one of the two measurement procedures.

4. Both uncertainties of X_A and X_B are again due to random fluctuations, and described by normal distributions, but now the discrepancy is smaller than the sum of the two "3σ" intervals (Fig. 4.11, lower left),

$$|X_A - X_B| < 3\sigma_A + 3\sigma_B . \qquad (4.34)$$

In this case, a decision about the consistency of the values X_A and X_B is less trivial than in the case of rectangular distributions. Only probabilistic considerations can be done, based on a comparison between the discrepancy and the width of the normal distributions (this topic is further considered in Sect. 9.3). The decision is to a good extent left to the subjective evaluation of the experimenter.

When dealing with inconsistent measures, before drawing definitive conclusions, it is good practice to carefully re-examine both procedures A and B, to try to find and eliminate the possible systematic errors. To this aim, if possible, one considers further reference measurements performed by different experimenters and by different procedures.

Example 4.18. Let us consider again the measurement of the elastic constant of a spring. The possible inconsistency between the results of the two measurement procedures, static and dynamic (Experiments E.3 and E.4 in Appendix E), can lead us to reconsider the dynamic procedure. In the relation $T = 2\pi(m/k)^{1/2}$, the mass m measures the inertia of the body attached to the spring, but does not take into account the inertia of the spring itself. A more complete relation is $T = 2\pi(M/k)^{1/2}$, where $M = m + m_e$, m_e being an effective mass that takes into account the spring inertia.

Weighted Average

Before going deeper into the treatment of systematic errors in the case of inconsistent measures, let us introduce here the procedure of weighted average, which is currently used to synthesize the results of two or more measurements repeated in different conditions.

Let us consider at first the case of two measures:

$$X_A \pm \delta X_A, \qquad X_B \pm \delta X_B. \qquad (4.35)$$

To calculate a unique average value X_0, it is reasonable to take into account the possible difference of uncertainties δX_A and δX_B: the smaller is the uncertainty of a value, the larger should be its contribution to the average. To this aim, the *weights* of the values X_A and X_B are defined as

$$w_A = \frac{1}{(\delta X_A)^2}, \qquad w_B = \frac{1}{(\delta X_B)^2}, \qquad (4.36)$$

and the value X_0 is calculated as a weighted average

$$X_0 = X_w = \frac{X_A w_A + X_B w_B}{w_A + w_B}. \qquad (4.37)$$

A formal foundation of the procedure of weighted average is given in Sect. 7.3. There, it is shown that the weighted average (4.37) is rigorously justified only when the uncertainties can be expressed as standard deviations of normal distributions. For other distributions, such as the rectangular distribution describing the measurement resolution, (4.37) can be used as an approximation, provided the uncertainty is expressed as the standard deviation of the distribution (say as a standard uncertainty).

The uncertainty δX_w of the weighted average of two consistent measures X_A and X_B, according to the procedures of propagation of uncertainty that are introduced in Chap. 8, is

$$\delta X_w = \frac{1}{\sqrt{w_A + w_B}}. \qquad (4.38)$$

Example 4.19. The elastic constant k of a spring is measured both statically and dynamically (Experiments E.3 and E.4 in Appendix E). The corresponding results are: $k_A = 10.40 \pm 0.04 \,\mathrm{kg\,s^{-2}}$ and $k_B = 10.37 \pm 0.08 \,\mathrm{kg\,s^{-2}}$. The weights of the two results are $w_A = 625 \,\mathrm{s^4\,kg^{-2}}$ and $w_B = 156 \,\mathrm{s^4\,kg^{-2}}$, respectively. The weighted average is $k = 10.39 \pm 0.03 \,\mathrm{kg\,s^{-2}}$.

The procedure of weighted average can be generalized to any number of measures:

$$X_w = \frac{\sum_i X_i w_i}{\sum_i w_i}, \qquad \text{where } w_i = \frac{1}{(\delta X_i)^2}, \qquad (4.39)$$

and the uncertainty, for consistent measures, is

$$\delta X_w = \frac{1}{\sqrt{\sum_i w_i}}. \qquad (4.40)$$

Uncertainty Due to Systematic Errors

Let us now consider two measures X_A and X_B that are inconsistent, due to the presence of nonnegligible systematic errors in at least one of the two procedures of measurement.

An *average value* can always be calculated as a weighted average (4.37), attributing a larger weight to the measure affected by the smallest uncertainty.

The discrepancy of the two measures can be connected to an uncertainty δX_{sys} due to systematic errors. The standard uncertainty due to systematic errors can be evaluated by the same procedure used for random fluctuations in (4.29), say as the estimated standard deviation of the distribution of sample means. For two values X_A and X_B, it is easy to verify, from (4.29), that the standard uncertainty is

$$\delta X_{\text{sys}} \simeq \frac{|X_A - X_B|}{2}. \tag{4.41}$$

This procedure can be generalized to any number of inconsistent measures $X_A, X_B, X_C, \ldots$, obtained in different conditions. The distribution of the values $X_A, X_B, X_C, \ldots$ is due to the different influence of systematic errors in each experiment. The average value X_0 can still be calculated by the weighted average (4.39). The uncertainty of the weighted average δX_w, calculated through (4.40), only takes into account the uncertainties of single measurements $X_A, X_B, X_C, \ldots$. The uncertainty δX_{sys} due to systematic errors can instead be evaluated, in analogy with (4.41), starting from the distribution of values, using the expression (4.29) introduced for random fluctuations.

Otherwise stated, for a large set of measurements performed in different conditions, the systematic errors of the single measures can be formally treated as random fluctuations; this procedure is sometimes referred to as *randomization of systematic errors*.

Example 4.20. In a university laboratory, N groups of students independently determine the gravitational acceleration g by measuring period T and length ℓ of a pendulum, and using the relation $T = 2\pi(\ell/g)^{1/2}$. Let $g_k \pm \delta g_k$ be the result of the kth group ($k = 1, 2, \ldots, N$). If the results of the different groups are inconsistent, the discrepancies should be attributed to systematic errors. Origin and extent of the systematic error of each group is unknown. However, if the number N of groups is large, it happens that the distribution of values tends to a normal shape. The uncertainty due to systematic errors can then be evaluated from the standard deviation of the distribution of values g_k.

The consistency or inconsistency of two or more measures $X_A, X_B, X_C, \ldots$ obtained in different conditions can be a posteriori evaluated by comparing the uncertainty δX_w of their weighted average (4.40) with the uncertainty calculated from the distribution of the single measures $X_A, X_B, X_C, \ldots$; if the latter is larger, the measures are inconsistent, due to the influence of systematic errors or to the underevaluation of uncertainties of single measures.

4.5 Summary and Comparisons

Three different causes of uncertainty have been distinguished in previous sections: the measurement resolution (Sect. 4.2), the dispersion of values due to random fluctuations (Sect. 4.3), and the estimate of the contributions of systematic errors (Sect. 4.4). Sometimes the effects of random fluctuations and systematic errors of instruments are evaluated by the manufacturer, and one has to take into account also the uncertainty quoted in the operation manual.

Uncertainty is an intrinsic part of the measure, and must always be explicitly quoted, for example, in the form $X_0 \pm \delta X$. The definition of the uncertainty δX has, however, a conventional character, and the very meaning of uncertainty can be different in different situations. Let us here consider two examples.

1. The uncertainty due to random fluctuations has been assumed equal to the standard deviation of the distribution of sample means, $\delta X = \sigma[m^*]$, but it could have been instead assumed $\delta X = 2\sigma[m^*]$ or $\delta X = 3\sigma[m^*]$.
2. The uncertainties due to random fluctuations and to resolution are connected to the width of two very different distributions, normal and rectangular, respectively. In the case of resolution, one can further distinguish between a maximum uncertainty $\Delta X/2$ and a standard uncertainty $\Delta X/\sqrt{12}$.

The lack of homogeneity in the definition and expression of uncertainty can cause nonnegligible difficulties in some common situations, such as the following.

(a) The uncertainty of a measure is due to the joint effect of different causes, such as random fluctuations and systematic errors.
(b) The uncertainty δQ of an indirectly measured quantity has to be expressed as a function of the uncertainties $\delta X, \delta Y, \ldots$, in general of different origin, of the directly measured quantities $X, Y, \ldots$
(c) One compares the measures of the same physical quantity obtained in different laboratories, which use different conventions to quote the uncertainties.

Following a suggestion of the B.I.P.M. (Bureau International des Poids et Mesures), in 1995 the I.S.O. (International Organization for Standardization) has issued some general recommendations for the evaluation and expression of uncertainty.

Statistical and Nonstatistical Uncertainty

The I.S.O. recommendations classify the uncertainties according to the methods of evaluation, rather than to their causes. Two general types of uncertainties are thus distinguished:

- Type A: uncertainties evaluated through statistical methods
- Type B: uncertainties evaluated through nonstatistical methods

Example 4.21. The uncertainty δX_{cas} due to random fluctuations is Type A, because its evaluation is based on the statistical treatment of a set of N repeated measurements.

Example 4.22. The uncertainty δX_{res} due to resolution is Type B, because its evaluation does not refer to statistical methods, but is based on the knowledge of the resolution interval ΔX.

Example 4.23. The uncertainty δX_{sys} due to the estimation of systematic errors is generally considered Type B, say of nonstatistical origin. However, when the estimate is based on the comparison between inconsistent measures, the uncertainty δX_{sys} should be considered Type A. It is convenient to associate with the systematic error a model distribution (normal, rectangle, triangle, etc.) whose shape depends on the available information.

Unified Expression of Uncertainty

The following conventions have been established in order to achieve a unified expression of uncertainty.

(a) With each cause of uncertainty, a suitable distribution is associated; the distribution can be of statistical origin (Type A) or a priori assumed on the base of available information (Type B).
(b) The uncertainty δX is assumed equal to the standard deviation of the distribution, and is called standard uncertainty.

Example 4.24. The uncertainty δX_{cas} due to random fluctuations is assumed equal to the standard deviation of the limiting distribution of sample means

$$\delta X_{\text{cas}} = \sigma[m^*], \qquad (4.42)$$

and can be estimated from experimental data through (4.29):

$$\delta X_{\text{cas}} = \sqrt{\frac{1}{N(N-1)} \sum_{i=1}^{N} (x_i - m^*)^2}. \qquad (4.43)$$

The distribution of sample averages is, to a good approximation, normal (Fig. 4.12, right); the interval from $X_0 - \sigma[m^*]$ to $X_0 + \sigma[m^*]$ contains 68% of the possible values.

Example 4.25. The uncertainty δX_{res} due to resolution is associated with a rectangular distribution, whose base and height are ΔX and $1/\Delta X$, respectively (Fig. 4.12, left). It is shown in Sect. 6.3 that the standard deviation of the rectangular distribution is $\sigma = \Delta X/\sqrt{12}$, so that

$$\delta X_{\text{res}} = \Delta X / \sqrt{12} . \tag{4.44}$$

The interval from $X_0 - \delta X$ to $X_0 + \delta X$ corresponds to 58% of the rectangle area.

Example 4.26. When dealing with systematic errors, the distribution is assumed on the base of available information. If we only know that the value X cannot be smaller than $X_{\min}$ and larger than $X_{\max}$, it is reasonable to assume a rectangular distribution (Fig. 4.12, left). If, however, there are good reasons to think that the central values are more probable than the values near the extrema $X_{\min}$ and $X_{\max}$, then it is more reasonable to assume a triangular or a normal distribution (Fig. 4.12, center and left). Once the distribution has been chosen, its standard deviation can be calculated, according to the rules that are introduced in Chap. 6, in order to determine δX_{sys}.

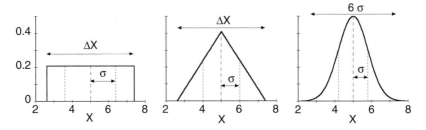

Fig. 4.12. Three different distributions, normalized to unit area: rectangular (left), triangular (center) and normal (right). To facilitate the comparison, the distributions have the same mean $m = 5$ and similar width. The standard deviation is $\sigma = \Delta X/\sqrt{12}$ for the rectangle and $\sigma = \Delta X/\sqrt{24}$ for the triangle; for the normal distribution, the interval of width 6σ centered on the mean includes 99.74% of the whole area. The percent area included between $X = m - \sigma$ and $X = m + \sigma$ is 58, 65, and 68% for the rectangular, triangular, and normal distributions, respectively.

There are many advantages in defining the uncertainty as the standard deviation of a probability distribution. In particular, it is possible to compare and, if necessary, coherently compose uncertainties of different origin. Moreover, many algorithms for statistical data analysis, such as the weighted average, are based on specific properties of the normal distribution and its parameters m and σ. If the uncertainty is always identified with a standard deviation, the algorithms developed for the normal distribution can be extended, to a good approximation, also to uncertainties of nonstatistical origin. This topic is further investigated in subsequent chapters.

4.5 Summary and Comparisons

The intervals defined by the standard deviation around the central value X_0 cover only part of the possible measurement values (about 68 and 58% for the normal and rectangular distributions, respectively). In some applications, typically referring to safety problems, one prefers to connect the uncertainty to an interval of values corresponding to a probability near to one (probability one corresponds to certainty). In these cases, one refers to an *extended uncertainty*, which corresponds to the standard uncertainty multiplied by a suitable *coverage factor* (Sect. 9.3), which is typically 2 or 3 for normal distributions. Nonstandard uncertainties do not comply with the I.S.O. recommendations; their use should be limited to cases of necessity and always accompanied by a warning.

Comparison Between the Different Causes of Uncertainty

The uncertainty due to resolution can always be evaluated. The uncertainty due to random fluctuations can be evaluated only if measurements are repeated. The two types of uncertainty are in general mutually exclusive and can be compared according to the following considerations.

Let us suppose that a physical quantity X is measured with a resolution ΔX, corresponding to an uncertainty $\delta X_{\text{res}} = \Delta X/\sqrt{12}$. Let us further suppose that random fluctuations give rise to a dispersion of values described by a normal distribution, with standard deviation σ_x; the corresponding uncertainty is $\delta X_{\text{cas}} = \sigma[m^*] = \sigma_x/\sqrt{N}$, where N is the number of measurements. When N increases, the uncertainty δX_{cas} decreases as $1/\sqrt{N}$. It is reasonable to reduce the uncertainty δX_{cas} by increasing the number N of measurements only as long as

$$\delta X_{\text{cas}} > \delta X_{\text{res}}, \quad \text{say} \quad \frac{\sigma_x}{\sqrt{N}} > \frac{\Delta X}{\sqrt{12}}. \tag{4.45}$$

The measurement resolution always represents a lower limit to uncertainty, $\delta X \geq \delta X_{\text{res}}$. It is good practice, when possible, to reduce the resolution, or limit the number N of measurements, so that $\delta X_{\text{cas}} > \delta X_{\text{res}}$, because the statistical treatment of random fluctuations, based on the normal distribution, is much better established.

It is relatively easy to evaluate, and in some cases to reduce, the uncertainties due to resolution and to random fluctuations. However, one should never neglect the third source of uncertainty, the presence of systematic errors, which can in some cases become predominant. Evaluating the uncertainty δX_{sys} due to systematic errors is in general not easy, and requires a critical and skillful analysis of the full measurement procedure. For that reason, it happens that δX_{sys} is often underevaluated or neglected completely. In some cases, however, as pointed out in Sect. 4.4, measurements repeated in different conditions allow a statistical evaluation of the influence of systematic errors.

Composition of Uncertainties

The uncertainty due to systematic errors is independent of the uncertainties due to resolution or random fluctuations. When quoting the uncertainty of a measure, it is useful to distinguish the two different contributions, δX_{sys} on the one hand, and δX_{cas} or δX_{res}. In some cases, different independent causes of systematic errors can be further distinguished, and it can be useful to separately quote the related uncertainties.

Sometimes, however, an overall expression of uncertainty δX_{tot} is sought, taking into account the contributions of the different causes (such as δX_{sys} and δX_{cas}, or δX_{sys} and δX_{res}). In these cases, if all uncertainties have been expressed as standard deviations of suitable distributions, the overall uncertainty has to be expressed as the quadratic sum of the component uncertainties. For example:

$$\delta X_{tot} = \sqrt{(\delta X_{cas})^2 + (\delta X_{sys})^2}, \qquad (4.46)$$

or, more generally, if δX_i are the different contributions to uncertainty,

$$\delta X_{tot} = \sqrt{\sum_i (\delta X_i)^2}. \qquad (4.47)$$

A justification of the quadratic sum procedure is given in Sect. 8.2.

Alternative Expressions of Uncertainty

Up to now the measures have been expressed in the form $X = X_0 \pm \delta X$, where the absolute uncertainty δX has the same dimension as the central value X_0.

The uncertainty δX should always be expressed by no more than two significant digits, a larger number of significant digits being generally meaningless. Correspondingly, the central value X_0 should be expressed by a number of significant digits consistent with the uncertainty (more details on significant digits are given in Appendix A.1).

An alternative expression of uncertainty is sometimes used, where the significant digits corresponding to the uncertainty δX are written in parentheses immediately after the central value X_0; the uncertainty is referred to the corresponding last significant digits of the central value X_0. For example

$$\ell = 2.357(25) \text{ m} \quad \text{corresponds to} \quad \ell = (2.357 \pm 0.025) \text{ m}.$$

This expression of uncertainty is useful when the uncertainty is much smaller than the central value, such as for fundamental constants of physics (Table C.2 in Appendix C). For example, the electron mass quoted as

$$m_e = 9.10938188(72) \times 10^{-31} \text{ kg}$$

corresponds to

$$m_e = (9.10938188 \pm 0.00000072) \times 10^{-31} \text{ kg}.$$

Relative Uncertainty

The quality of a measurement cannot be solely determined by its absolute uncertainty δX. For example, an absolute uncertainty $\delta X = 1\,\text{mm}$ has different meaning when referring to a length $X_0 = 1\,\text{cm}$ or to a length $X_0 = 10\,\text{m}$. The quality of a measurement is better expressed by the relative uncertainty

$$\frac{\delta X}{|X_0|} . \tag{4.48}$$

The smaller the relative uncertainty is, the higher the quality of the measurement. The modulus $|X_0|$ in (4.48) gives a positive value of relative uncertainty even if X_0 is negative. By definition, the relative uncertainty is always a dimensionless quantity.

The relative uncertainty can be much smaller than one. To avoid the use of many decimal digits, the relative uncertainty is sometimes multiplied by 100, and is called *percent uncertainty*, indicated by %.

If the absolute uncertainty is many orders of magnitude smaller than the central value, as happens for many fundamental constants of physics (Table C.2 in Appendix C), the relative uncertainty is multiplied by 10^6, and expressed in *parts per million* (ppm).

Example 4.27. The length of a pendulum, $\ell = 1.25\,\text{m}$, and the elongation of a spring, $x = 1.2\,\text{cm}$, are measured with the same absolute uncertainty $\delta\ell = \delta x = 1\,\text{mm}$. The relative uncertainty $\delta\ell/\ell = 8 \times 10^{-4} = 0.08\%$ in the former case is much smaller than $\delta x/x = 8 \times 10^{-2} = 8\%$ in the latter case.

Example 4.28. The electron mass $m_e = 9.10938188(72) \times 10^{-31}\,\text{kg}$ has a relative uncertainty $\delta m/m = 8 \times 10^{-8}$, corresponding to 0.08 ppm.

Qualitative Characteristics of Measures

Uncertainty is a quantitative property of a measure, and is expressed by a numerical value.

Other terms are often used to qualitatively characterize the results of measurements. The following definitions are consistent with the vocabulary of metrology terms proposed by a task group of I.S.O. in 1993.

By *repeatability* one qualifies the closeness of agreement between the results of successive measurements of the same physical quantity carried out under the same conditions (procedure, observer, instruments, and location) and repeated over a short time interval. Repeatability is connected to random fluctuations.

By *reproducibility* one qualifies the closeness of agreement between the results of measurements of the same physical quantity carried out under changed conditions (procedures, experimenters, instruments, sites, and times).

74   4 Uncertainty in Direct Measurements

By *accuracy* one qualifies the closeness of the agreement of a measure and a true value of a physical quantity. A *true value* means here a value that is accepted, sometimes by convention, as having an uncertainty suitable for a given application. For example, for many didactic applications, one can consider as true the values of the fundamental constants of physics periodically published by the international group CODATA (Table C.2 in Appendix C).

Problems

4.1. Calculate the variance D^* and the standard deviation σ^* of the following sets of six numbers.

(a) 4, 4, 4, 8, 8, 8
(b) 2, 5, 6, 6, 7, 10
(c) 3, 6, 6, 6, 6, 9

For each set of numbers, draw the histogram and compare the difference between the maximum and minimum values, $\Delta X = X_{max} - X_{min}$, with twice the standard deviation $2\sigma^*$.

4.2. The length of $N = 20$ bolts has been measured by a wernier caliper with resolution 0.05 mm (see Experiment E.1 of Appendix E). The following values (in millimeters) have been obtained.

20.00 20.00 19.90 19.95 20.15 19.90 20.05 20.00 20.10 20.10
20.10 20.05 19.90 19.90 20.10 20.05 20.10 19.90 20.00 20.05

Represent the values in a handmade histogram, and calculate mean, variance, and standard deviation. Plot the height-normalized and area-normalized histograms.

Notice that here the histogram represents the values of 20 different quantities (the lengths of the 20 bolts), not 20 measures of the same quantity.

4.3. The period of a pendulum has been measured 200 times by a stopwatch with resolution 0.01 s (Experiment E.2 of Appendix E). The following table lists the different values T_i of the period (top row) and the corresponding number n_i of measures (bottom row).

T_i	1.91	1.92	1.93	1.94	1.95	1.96	1.97	1.98	1.99	2.00	2.01	2.02	2.03	2.04	2.05
n_i	4	4	2	8	15	19	18	19	56	16	10	19	6	3	1

Plot the height-normalized and area-normalized histograms. Calculate sample mean, sample variance and sample standard deviation.

Estimate the mean and variance of the limiting normal distribution (4.17). Plot the normal distribution in the same graph as the area-normalized histogram.

Estimate the variance of the distribution of sample means. Evaluate the uncertainty due to random fluctuations and compare it with the uncertainty due to resolution. Express the value of the period as $T_0 \pm \delta T$.

4.4. Plot on the same graph the normal distributions (4.17) with $m = 0$ and $\sigma = 1, 2, 3, 4$.

4.5. Evaluate and compare the relative uncertainties of the fundamental constants listed in Table C.2 of Appendix C.

Part II

Probability and Statistics

5 Basic Probability Concepts

The uncertainty of a measure is related to the width of the probabilistic distribution of its possible values, such as a normal distribution, a rectangular distribution, or a triangular distribution.

In this chapter, some basic concepts of probability theory are introduced; they represent a necessary background for the more specific study of the distributions of random variables and statistical techniques, to which the subsequent Chaps. 6 and 7 are dedicated, respectively.

5.1 Random Phenomena

The casual dispersion of the results of repeated measurements (Sect. 4.3) is a typical example of a *random phenomenon*. In general, a phenomenon is said to be *random* or *casual* if, when reproduced with the same initial conditions, it goes on with unpredictable outcomes.

Example 5.1. A die is cast many times in the same way. Its trajectory is casual, and the outcome is unpredictable.

In many physical phenomena, one can single out some central characteristics that can be described by deterministic laws: once the initial conditions are known, the development and the final outcome can be precisely predicted. There are, however, also some secondary characteristics that are impossible, or inconvenient, to treat according to deterministic laws, and that give rise to a random behavior; their influence can be of different extent, but is never completely removable.

Example 5.2. The trajectory of a projectile that is thrown at an angle θ with respect to the horizontal, with initial velocity v_0, can be exactly determined by the laws of mechanics, once the acceleration of gravity g is known; for example, the range is $s = v_0^2 \sin 2\theta / g$. The real trajectory will, however, be influenced by random secondary factors, such as vibrations, winds, and laying errors.

Example 5.3. The measurement of a physical quantity is, to a large extent, a deterministic process. It is, however, impossible to completely eliminate the influence of casual fluctuations, and the result of a measurement is an interval of values instead of a single value.

Some phenomena exist, for which a deterministic treatment is altogether impossible. In some cases, this is merely due to the complexity of the phenomenon; in other cases, typically in atomic and subatomic physics, the phenomenon has a genuine random nature.

Example 5.4. Let us consider a macroscopic volume of gas, say $1\,\mathrm{dm}^3$. The trajectory of one molecule could in principle be described by the laws of classical mechanics. However, the exceedingly large number of molecules, typically of the order of 10^{23}, makes a deterministic treatment of their global behavior impossible. An effective treatment is based on considering the behavior of each molecule as perfectly random.

Example 5.5. The decay of a radioactive isotope is a genuinely random phenomenon, that by no means can be deterministically predicted.

Finally, some phenomena are expressly designed in order to exhibit random characteristics.

Example 5.6. Games of chance are typically based on phenomena whose outcome is deterministically unpredictable, such as dice tossing or card playing.

Example 5.7. Complex physical systems can sometimes be effectively studied by computer simulation of their behavior. Some methods are based on the perfect randomness of the elementary choices performed by the computer at each simulation step.

The unpredictability, the complexity, and the large number of causes that characterize random phenomena require specific methods for their treatment. Such methods are based on the theory of probability and on statistics. The possibility of using the rigorous mathematical methods of probability theory to describe random phenomena is guaranteed by the following well-established experimental evidence. The outcome of a single random phenomenon is completely unpredictable; the repetition of the phenomenon will give rise to a random distribution of outcomes. When, however, the number of repetitions becomes very large, the outcome distribution assumes characteristics of increasing regularity, and the average values of some relevant quantities become increasingly stable.

Example 5.8. The toss of a coin is a random phenomenon: the outcome, head or tail, is unpredictable. However, when the number of tosses increases, the frequency of heads (as well as of tails) tends to stabilize around the value $1/2$.

Example 5.9. In a volume of gas in thermodynamic equilibrium, the collisions of the molecules with the walls of the container are randomly distributed in space and time. However, if the number of molecules is sufficiently large, their effect appears as a constant and uniform pressure.

Example 5.10. The decay of a single radioactive isotope is a genuinely random phenomenon. However, if a very large number of isotopes is considered, the time sequence of their decays exhibits a well-defined exponential behavior.

Example 5.11. A single measurement of a physical quantity affected by random fluctuations gives a casual value. However, if the measurement is repeated a large enough number of times, the distribution of measures tends to assume a well-defined shape that can be synthesized by two numerical parameters, the mean and the standard deviation.

5.2 Sample Space. Events

A random phenomenon can have different outcomes. The set S whose elements are all the possible outcomes of a random phenomenon is called the *sample space*. The sample space can have a finite, countable infinite or uncountable infinite number of elements.

Example 5.12. By tossing a die, one can obtain six outcomes: 1, 2, 3, 4, 5, 6. The sample space is $S = \{1, 2, 3, 4, 5, 6\}$. The number of elements of the sample space is finite.

Example 5.13. A coin is tossed two times. Four outcomes are possible: HT, TH, HH, TT (H = head, T = tail). The sample space is $S = \{$HT, TH, HH, TT$\}$. The number of elements of the sample space is finite.

Example 5.14. An experiment consists of repeatedly tossing a coin, until "head" is for the first time obtained. The possible outcomes are the elements of the sample space $S = \{H, TH, TTH, TTTH, TTTTH, \ldots\}$, that is in biunivocal correspondence with the set of natural numbers $\{1, 2, 3, 4, \ldots\}$, and has then a countable infinite number of elements.

Example 5.15. The breakdown of the filament of an incandescent lamp is a random phenomenon, and the lifetime is unpredictable. The sample space is represented by all possible intervals of time $\Delta t \geq 0$, and has an uncountable infinite number of elements.

Any possible subset A of the sample space S is an *event*: otherwise stated, an event is a set of possible outcomes (Fig. 5.1, left). An event A is realized if one of the outcomes belonging to it is realized. The single outcomes are sometimes called simple events.

Example 5.16. A coin is tossed twice. The sample space is $S = \{$TT, TH, HT, HH$\}$. The event "one head" is the subset $A = \{TH, HT\}$ (Fig. 5.1, right).

Example 5.17. A die is tossed ten times. An outcome is a sequence of ten numbers, included between 1 and 6. There are 6^{10} possible different sequences, the sample space S has thus 6^{10} elements. Possible events are "6 appears 5 times", "6 appears 3 times", "3 appears 2 times and 2 appears 4 times", and "4 appears 3 times consecutively".

Fig. 5.1. Left: an event A is a subset of the sample space S. Right: sample space for two tosses of a coin; event $A =$ "head appears once".

Example 5.18. Let us consider the lifetime of 1000 incandescent lamps. The sample space is the product of the sample spaces of each lamp, and every element is represented by 1000 lifetime values. Possible events are "100 lamps have a lifetime longer than 4000 hours", "no lamp has a lifetime shorter than 10 hours", "20 lamps break down within the first 10 hours".

It is convenient to single out two particular events:

- The *certain event* $A = S$ (A coincides with the sample space).
- The *impossible event* $A = \emptyset$ ($\emptyset$ is the empty set).

Example 5.19. A die is tossed. The event "the outcome is a number larger than 0 and smaller than 7" is certain. The event "the outcome is 8" is impossible.

5.3 Probability of an Event

The outcome of a single random phenomenon cannot be predicted in a deterministic way. Common experience, however, shows that different events related to the same phenomenon can happen more or less easily.

Example 5.20. A coin is tossed ten times. We expect that the realization of the event "the outcomes are 6 tails and 4 heads" is easier than the realization of the event "the outcomes are 10 heads".

To quantitatively express the ease of realization, one associates with every event a real number $\mathcal{P}$, that is called the *probability of the event*. By convention, the extremum values of $\mathcal{P}$ are:

$$\mathcal{P} = 1 \quad \text{for the certain event.}$$
$$\mathcal{P} = 0 \quad \text{for the impossible event.}$$

The probability of an event is thus a number included between 0 and 1 ($0 \leq \mathcal{P} \leq 1$).

Note 1. The possibility of attributing a probability to every event is relatively easy if the outcomes are a discrete set; if the outcomes are a continuous set, a less intuitive approach is necessary, that is considered in Chap. 6.

Note 2. If an event is impossible, its probability is by definition $\mathcal{P} = 0$. On the contrary, an event with probability $\mathcal{P} = 0$ is not necessarily impossible: this is the case of continuous events that are treated in Chap. 6.

The problem of attributing a reasonable value of probability $\mathcal{P}$ to random events has been solved through several different empirical rules, which are considered below in this section. These rules allow the evaluation of the probability of simple events. Once the probabilities of simple events are known, the probabilities of complex events can be rigorously calculated by the methods of the theory of probability, whose basic concepts are considered shortly in the next sections. The theory of probability is based on an axiomatic definition of probability, that is introduced in Sect. 5.5.

Classical Probability

The first rule for calculating the probability of an event was introduced by P. S. Laplace (1749–1828). According to this rule, the probability of an event is evaluated "a priori", without performing experiments, as the ratio between the number of outcomes that correspond to the event and the total number of possible outcomes in the sample space:

$$\mathcal{P}(A) = \frac{\text{number of favorable outcomes}}{\text{total number of outcomes}} = \frac{m}{M}. \qquad (5.1)$$

Example 5.21. A die is tossed. The probability that the outcome is "3" is $\mathcal{P}(3) = 1/6$. In fact, there are $M = 6$ possible outcomes, $m = 1$ of which is favorable.

Example 5.22. A box contains 5 marbles, 2 white and 3 brown. One marble is drawn: the probability that it is white is $\mathcal{P}(\text{white}) = 2/5$. In fact, there are $M = 5$ possible outcomes, $m = 2$ of which are favorable.

The classical rule can be used only if the following requirements are fulfilled.

1. Of all possible outcomes, one and only one is realized.
2. The possible outcomes have equal likelihood: a given outcome has no particular advantages with respect to any other outcome.

These requirements can be considered a priori fulfilled only for phenomena affected by peculiar symmetries. The classical rule is actually inadequate for many scientific and technological problems. In addition, the requirement of equal likelihood of the outcomes implies a certain circularity of the classical rule.

Statistical Probability

An alternative rule for attributing a probability to a random event, with a larger range of applications, is connected to the concept of *statistical frequency*. This rule is called "a posteriori", because it requires the previous performance of a number of experiments. Let us suppose that an experiment is repeated N times, and the event A is realized in n^* of the N experiments ($n^* \leq N$); the statistical frequency (or sample frequency) of the event A relative to the N experiments is the ratio:

$$p^*(A) = \frac{n^*}{N}. \tag{5.2}$$

The statistical frequency has a random character. If the set of N experiments is repeated many times, the frequency p^* assumes different values at each repetition, in an unpredictable way.

Example 5.23. A coin is tossed 100 times, and the statistical frequency of "head" is $p_1^* = 45/100$. The same coin is again tossed 100 times, and now the statistical frequency is $p_2^* = 52/100$. And so on.

It is, however, a matter of experience that the relative amplitude of the random fluctuations of the values p^* of statistical frequency tends to decrease when the number N of experiments increases. Otherwise stated, when N increases, the value of p^* tends to stabilize. For the cases where the classical rule (5.1) can be used (such as tossing of dice), one can experimentally verify that the statistical frequency tends to the probability value of the classical rule.

Following these considerations, R. von Mises (1883–1953) proposed considering the probability $\mathcal{P}(A)$ of an event A as the limiting value of the statistical frequency $p^*(A) = n^*/N$, for $N \to \infty$. This proposal does not correspond to an operative definition of probability, because N is necessarily finite in a real experiment. Anyway, it is common practice to express the probability as

$$\mathcal{P}(A) \simeq p^*(A) = \frac{n^*}{N}, \tag{5.3}$$

with the understanding that the larger N is, the better is the probability evaluated through (5.3).

Subjective Probability

Both rules, the classical one (5.1) and the statistical one (5.3), are affected by logical incoherences and practical difficulties. A tentative solution of these difficulties is represented by *subjective probability* (F. P. Ramsay, 1903–1930; B. de Finetti, 1906–1985), that is, a measure of the degree of confidence that a given person has in the realization of a given event. Obviously, the subjective evaluation has always to take into account, although not in a deterministic way, the results of both classical and statistical rules.

Example 5.24. A person attributes the probability $\mathcal{P} = 1/6$ to the outcome "5" when a die is tossed, on the basis of symmetry considerations and of the experience gained in previous repeated experiments. He thus expects that the frequency of "5" approaches 1/6 when the number of tosses increases.

Examples

The meaning of the different rules for calculating the probability can be better grasped by considering some examples.

Example 5.25. A village has $M = 2000$ inhabitants, of whom $m_1 = 500$ are younger than 20, $m_2 = 1100$ are between 20 and 60, and $m_3 = 400$ are older than 60. One name is chosen by chance from the list of the inhabitants. What is the probability that the chosen person is younger than 20? We can use here the classical rule, because the outcomes are equivalent and mutually exclusive. One gets $\mathcal{P} = m_1/M = 500/2000 = 1/4$.

Example 5.26. A box contains 2 white and 3 brown marbles. Two marbles are drawn by chance. We seek the probability that both marbles are brown. Three different procedures of extraction can be envisaged: (a) the first marble is repositioned in the box before drawing the second one; (b) the two marbles are drawn in sequence, without repositioning; (c) the two marbles are contemporarily drawn. The sample space can be built up for the three procedures (Fig. 5.2) and the classical rule can be used. The probability that both marbles are brown is:

(a) $\mathcal{P} = 9/25$ for the drawing with reposition.
(b) $\mathcal{P} = 3/10$ for the drawing in sequence without reposition.
(c) $\mathcal{P} = 3/10$ for the simultaneous drawing.

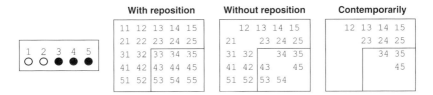

Fig. 5.2. Illustrating Example 5.26. The box with the five marbles (left), and the three sample spaces relative to the three different drawing procedures.

Example 5.27. Mr. White and Mr. Brown agree to meet at a given place between 11 and 12 a.m. Each of them will arrive at a random time between 11 and 12, and will wait exactly 20 minutes. What is the probability that Mr. White and Mr. Brown succeed in meeting? The sample space S is represented by all possible pairs of arrival times of Mr. White and Mr. Brown

(t_W, t_B); the event A is the subset of the pairs that satisfy $|t_W - t_B| \leq 20$ minutes. Both sets S and A have an uncountable infinite number of elements. The probability of meeting can be calculated by the classical rule through a geometrical argument. Let us consider a Cartesian plane, and represent the time of arrival of Mr. White and Mr. Brown on the horizontal and vertical axes, respectively (Fig. 5.3). The sample space S is represented by a square whose basis is $\Delta t = 60$ minutes, and the event A is represented by the grey zone. The probability is the ratio between the two surfaces: $\mathcal{P} = 5/9$.

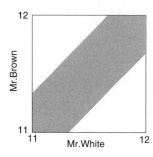

Fig. 5.3. Illustrating Example 5.27. On the horizontal and vertical axes, the times of arrival of Mr. White and Mr. Brown, respectively.

Example 5.28. The period $\mathcal{T}$ of a pendulum is measured N times. The results are shown in a histogram with column width $\Delta \mathcal{T} = 0.1\,\mathrm{s}$ (Fig. 5.4, left and center). One wants to evaluate the probability that the result of a further measurement is included between the values 2.2 and 2.3 s. The classical rule cannot be used here, and one has to rely on the statistical rule. From the histogram, one can calculate the statistical frequency of the values between 2.2 and 2.3 s, $p^* = 0.223$. To evaluate the probability, one assumes $\mathcal{P} = p^* = 0.223$. A more accurate evaluation of probability is based on the hypothesis that the limiting distribution is normal (Fig. 5.4, right); mean value m and standard deviation σ of the normal distribution can be estimated from the experimental values (Sect. 4.3). The probability $\mathcal{P}$ is the area under the normal distribution corresponding to the interval from 2.2 to 2.3 s; in this case, $\mathcal{P} = 0.241$. The hypothesis that the limiting distribution is normal depends, to a certain extent, also on subjective evaluations.

Example 5.29. In a laboratory report, the effect of systematic errors on the result of a measurement is quoted by simply assessing that the result is included between the values $X_{\min}$ and $X_{\max}$; no further information is given on the shape of the distribution of the values X. A reader of the report assumes that all values within the interval between $X_{\min}$ and $X_{\max}$ have the same probability. This is a typical example of subjective probability.

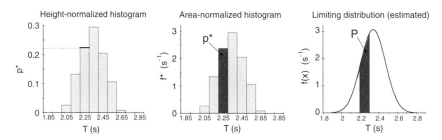

Fig. 5.4. Illustrating Example 5.28. Left: height-normalized experimental histogram; the heights of the columns correspond to the statistical frequency. Center: the same histogram, area-normalized; the statistical frequency now corresponds to the area of the columns. Right: normal distribution bestfitting the experimental histogram.

5.4 Addition and Multiplication of Events

Two basic operations can be defined on the sample space, addition and multiplication of events.

Addition of Events

Let us consider two events A and B. A third event C is said to be the sum of the events A and B, if C occurs whenever A or B occurs. The event C is the subset of the sample space S, that contains the elements of both the A and B events. In the language of set theory, C is the union of the A and B subsets (Fig. 5.5, a, b):

$$C = A + B \quad \text{or} \quad C = A \cup B . \tag{5.4}$$

Similarly, one defines the sum of more than two events:

$$C = A_1 + A_2 + A_3 + \cdots + A_n . \tag{5.5}$$

Example 5.30. A card is chosen from a deck of 52 cards. If $A =$ "the card is heart", $B =$ "the card is diamond", then $C = A + B =$ "the card is a red suit".

Example 5.31. Let us consider the lifetime of an electronic device. If $A =$ "the lifetime is included between 500 and 2000 hours" and $B =$ "the lifetime is included between 1000 and 4000 hours", then $C = A + B =$ "the lifetime is included between 500 and 4000 hours".

Multiplication of Events

Let us consider two events A and B. A third event C is said to be the product of the events A and B, if C occurs whenever both A and B occur. The event

C is the subset of the sample space S whose elements belong to both the A and B events. In the language of set theory, C is the intersection of the A and B subsets (Fig. 5.5, c):

$$C = AB \quad \text{or} \quad C = A \cap B. \tag{5.6}$$

Similarly, one can define the product of more than two events:

$$C = A_1 \, A_2 \, A_3 \cdots A_n. \tag{5.7}$$

Example 5.32. A card is chosen from a deck of 52 cards. If A = "the card is heart" and B = "the card is a face card", then $C = A \, B$ = "the card is a heart face card".

Example 5.33. Two dice are tossed. The elements of the sample space S are the 36 possible outcomes (pairs of numbers). The event A = "the sum is 8" corresponds to 5 outcomes. The event B = "two even numbers appear" corresponds to 9 outcomes. The event $C = AB$ = "two even numbers appear whose sum is 8" corresponds to 3 outcomes.

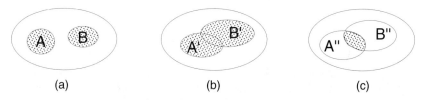

Fig. 5.5. In (a) and (b), the dotted area is the sum of events $A + B$ and $A' + B'$, respectively. In (c), the dotted area is the product of the events A'' and B''.

Mutually Exclusive Events

Two events A and B are said to be mutually exclusive if they cannot occur contemporarily. In the language of set theory, two events A and B are mutually exclusive if their intersection is the empty set (Fig. 5.6):

$$A \cap B = \emptyset. \tag{5.8}$$

Example 5.34. A card is chosen from a deck of 52 cards. The event A = "the card is heart" and the event B = "the card is a black suit" are mutually exclusive.

Contrary Events

An event $\overline{A}$ is said to be contrary to an event A, when $\overline{A}$ occurs if and only if A does not occur. In the language of set theory (Fig. 5.6, b):

$$A \cap \overline{A} = \emptyset, \quad A \cup \overline{A} = S. \tag{5.9}$$

Example 5.35. A die is tossed. If A = "an even number appears", then $\overline{A}$ = "an odd number appears".

(a)   (b)

Fig. 5.6. The events A and B in (a), as well as the events A and $\overline{A}$ in (b), are mutually exclusive. The event A in (b) is contrary to the event $\overline{A}$, and viceversa.

Complex Events

The operations of addition and multiplication of events, as well as the concept of contrary events, are useful to decompose complex events into sums and products of simpler events.

As an example, let us repeat the same trial three times, such as firing three shots at a target, and suppose that the outcome of each trial is independent of the outcomes of the other two. The sample space is

$$S = \{123, 12\overline{3}, 1\overline{2}3, \overline{1}23, 1\overline{2}\overline{3}, \overline{1}2\overline{3}, \overline{1}\overline{2}3, \overline{1}\overline{2}\overline{3}\}, \tag{5.10}$$

where $1, 2, 3$ label the positive outcomes, and $\overline{1}, \overline{2}, \overline{3}$ label the negative outcomes.

Working on the sample space S can be of little use when there is no certainty that its elements are equally probable. A more effective approach consists of separately considering each trial. Let A_s be the event "the sth trial has positive outcome", and $\overline{A}_s$ the contrary event "the sth trial has negative outcome". The event B = "only one of the three trials is positive" can be decomposed into sums and products

$$B = A_1 \overline{A}_2 \overline{A}_3 + \overline{A}_1 A_2 \overline{A}_3 + \overline{A}_1 \overline{A}_2 A_3. \tag{5.11}$$

As shown in the next sections, probability theory gives the rules for calculating the probability of complex events, once the probability of simpler events is known.

5.5 Probability of the Sum of Events

Let us consider the following problem. If the probability of two events A and B is known, how can the probability of the sum $A + B$ be evaluated?

Sum of Mutually Exclusive Events

If the two events A and B are mutually exclusive ($A \cap B = \emptyset$), the probability of their sum $A + B$ is the sum of their probabilities (Fig. 5.7 a):

$$\mathcal{P}(A + B) = \mathcal{P}(A) + \mathcal{P}(B) . \tag{5.12}$$

Eq. (5.12) can be easily justified in terms of classical probability (5.1):

$$\mathcal{P}(A) = \frac{m_A}{M}, \quad \mathcal{P}(B) = \frac{m_B}{M} \quad \Rightarrow \quad \mathcal{P}(A + B) = \frac{m_A + m_B}{M} . \tag{5.13}$$

When the classical rule cannot be used to calculate the probability of the events A and B, (5.12) is considered a priori valid, and is considered one of the postulates of the axiomatic theory of probability (see below).

The generalization of (5.12) to the sum of many mutually exclusive events is straightforward:

$$\mathcal{P}(A + B + C + \cdots + Z) = \mathcal{P}(A) + \mathcal{P}(B) + \cdots + \mathcal{P}(Z) . \tag{5.14}$$

Moreover, if $\overline{A}$ is the contrary of A, then

$$\mathcal{P}(\overline{A}) + \mathcal{P}(A) = \mathcal{P}(\overline{A} + A) = 1 , \quad \mathcal{P}(\overline{A}) = 1 - \mathcal{P}(A) . \tag{5.15}$$

Example 5.36. The probability that a lamp breaks down within the first 1000 hours is $\mathcal{P}(A_1) = 0.20$; the probability that it works more than 1000 hours and less than 2000 hours is $\mathcal{P}(A_2) = 0.25$. The probability that the lamp works more than 2000 hours is $\mathcal{P}(\overline{A}) = 1 - \mathcal{P}(A) = 1 - \mathcal{P}(A_1 + A_2) = 1 - \mathcal{P}(A_1) - \mathcal{P}(A_2) = 0.55$

General Rule for the Sum

If the two events are not mutually exclusive (say $A \cap B \neq \emptyset$) (Fig. 5.7 b), the probability of their sum is

$$\mathcal{P}(A + B) = \mathcal{P}(A) + \mathcal{P}(B) - \mathcal{P}(AB) . \tag{5.16}$$

The general rule (5.16) contains (5.12) as a particular case, inasmuch as $\mathcal{P}(AB) = 0$ for mutually exclusive events.

For three events A, B, C (Fig. 5.7 c), the probability of the sum is

$$\begin{aligned}\mathcal{P}(A + B + C) = \;&\mathcal{P}(A) + \mathcal{P}(B) + \mathcal{P}(C) \\&- \mathcal{P}(AB) - \mathcal{P}(AC) - \mathcal{P}(BC) \\&+ \mathcal{P}(ABC).\end{aligned} \tag{5.17}$$

The generalization to more than three events is straightforward.

To evaluate (5.16) and (5.17), it is necessary to know how to calculate the probability of the product of events, such as $\mathcal{P}(AB)$ and $\mathcal{P}(ABC)$. This problem is considered in Sect. 5.6.

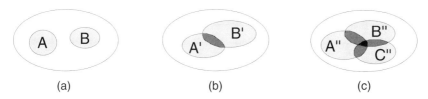

Fig. 5.7. In (a), the events A and B are mutually exclusive: $\mathcal{P}(A+B) = \mathcal{P}(A) + \mathcal{P}(B)$. In (b), the events A' and B' are not mutually exclusive; if the probabilities $\mathcal{P}(A')$ and $\mathcal{P}(B')$ are simply summed, the grey area is counted twice. In (c), if the probabilities $\mathcal{P}(A'')$, $\mathcal{P}(B'')$, and $\mathcal{P}(C'')$ are simply summed, the grey area is counted twice, and the black area is counted three times.

Axiomatic Definition of Probability

In the axiomatic constructions of the theory of probability, the probability is defined as a real number $\mathcal{P}$ that satisfies the following axioms.

(a) For any event A, $\mathcal{P}(A) \geq 0$.
(b) For the certain event S, $\mathcal{P}(S) = 1$.
(c) For any pair A, B of mutually exclusive events,

$$\mathcal{P}(A + B) = \mathcal{P}(A) + \mathcal{P}(B).$$

These axioms are the basis for demonstrating the theorems of the theory of probability, which allow one to calculate the probability of complex events from the probabilities of simpler events. The probabilities of the simple events always have to be evaluated by the empirical rules of Sect. 5.3.

5.6 Probability of the Product of Events

Let us now consider the following problem. If the probability of two events A and B is known, how can the probability of their product AB be evaluated? In order that the product event AB be realized, it is necessary that both events A and B are realized. It is thus necessary to verify if and how the realization of one of the two events, A or B, influences the realization of the other event, B or A, respectively.

Conditional Probability

The conditional probability $\mathcal{P}(B|A)$ of an event B, given A, is the probability of realization of B under the condition that A is realized. The conditional probability of B given A is calculated by taking, as a sample space of B, the set of the outcomes that realize A.

Example 5.37. Let us consider the toss of a die, and focus on the two events: A = "the outcome is an even number", B = "the outcome is a number less than 6". By considering the sample space (Fig. 5.8), one can separately calculate the probability of the two events: $P(A) = 1/2$ and $P(B) = 5/6$, as well as the probability of their product: $P(AB) = 1/3$. The conditional probability of A given B is $P(A|B) = 2/5$, whereas the conditional probability of B given A is $P(B|A) = 2/3$.

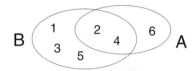

Fig. 5.8. Sample space of Example 5.37.

Example 5.38. Let us consider again Example 5.26 of Sect. 5.3. A box contains 2 white and 3 brown marbles (Fig. 5.9, left). Two marbles are drawn by chance in sequence, without reposition. Let us focus on the two events: A = "the first marble is brown"; B = "the second marble is brown". By considering the sample space (Fig. 5.9, center and right), one can easily see that $P(A) = P(B) = 3/5$. The conditional probability of B given A is $P(B|A) = 1/2$; also the conditional probability of A given B is $P(A|B) = 1/2$.

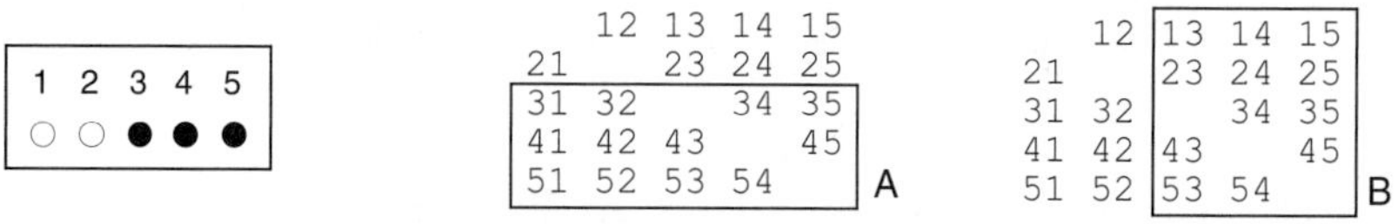

```
                    12 13 14 15              12 13 14 15
              21       23 24 25         21      23 24 25
1 2 3 4 5     31 32       34 35         31 32      34 35
○ ○ ● ● ●     41 42 43       45         41 42 43      45
              51 52 53 54       A       51 52 53 54      B
```

Fig. 5.9. Illustrating Example 5.38. On the left, the box containing five marbles. On the center and on the right, the sample space, where the boxes represent the subsets corresponding to the events A and B, respectively.

One can easily verify that the symmetric relations

$$P(A|B) = P(AB)/P(B) , \quad P(B|A) = P(AB)/P(A) \qquad (5.18)$$

are satisfied in the two examples 5.37 and 5.38. One can demonstrate that the equalities (5.18) have general validity.

Probability of the Product

Making use of the equalities (5.18), the probability $P(AB)$ of the product of two events A and B can be calculated as the product of the probability of one of the events by the conditional probability of the other event:

$$P(AB) = P(A)\,P(B|A) = P(B)\,P(A|B)\,. \tag{5.19}$$

Example 5.39. Let us consider again Example 5.37. A die is tossed, and let us focus on the two events: A = "the outcome is even", B = "the outcome is less than 6". The probability of the product AB = "the outcome is even and less than 6" is $P(AB) = P(A)\,P(B|A) = P(B)\,P(A|B) = 1/3$.

Example 5.40. Let us consider again Example 5.38. Two marbles are drawn in sequence without reposition from a box containing 2 white and 3 brown marbles, and let us focus on the two events: A = "the first marble is brown"; B = "the second marble is brown". The probability of the product AB = "both marbles are brown" is $P(AB) = P(A)\,P(B|A) = P(B)\,P(A|B) = 3/10$.

The probability of the product of more than two events is

$$P(A_1 A_2 \cdots A_n) = P(A_1)\,P(A_2|A_1)\,P(A_3|A_2 A_1) \cdots P(A_n|A_1 A_2 A_3 \cdots A_{n-1}). \tag{5.20}$$

Example 5.41. Five cards are chosen from a deck of 52. Let A be the event "the ith card is a diamond". The probability that all five cards are diamonds is $P(A_1 A_2 A_3 A_4 A_5) = (13/52)\,(12/51)\,(11/50)\,(10/49)\,(9/48) = 4.9 \times 10^{-4}$.

Independent Events

An event B is said to be *independent* of an event A, if the probability that B is realized is independent of the realization of A. If B is independent of A, A is independent of B as well.

Example 5.42. A die is tossed twice. The outcome of each toss is independent of the outcome of the other.

Example 5.43. A card is drawn from a deck of 52. The probability of it being an ace (event B) is independent of the probability of it being a diamond (event A).

Example 5.44. Two cards are drawn from a deck of 52. The probability that the second card is an ace (event B) is different according to whether the first card is an ace (event A) or not (event $\overline{A}$). The events A and B are not independent.

Example 5.45. Two marbles are drawn from a box containing two white and three brown marbles. Let us focus on the two events: A = "the first marble is brown"; B = "the second marble is brown". If the first marble is repositioned inside the box before drawing the second one, the two events A and B are independent. If, on the contrary, the first marble is kept outside the box, the events A and B are not independent.

The last examples can be generalized. The drawings of two or more objects out of a given set can be done with or without repositioning each object before drawing the next one. The drawing with and without repositioning are examples of independent and not independent events, respectively.

For independent events, the conditional probability reduces to the simple probability:

$$P(A|B) = P(A), \quad P(B|A) = P(B) \quad \text{(independent events)}, \quad (5.21)$$

and the probability of the product (5.19) reduces to

$$P(AB) = P(A) \, P(B) \quad \text{(independent events)}. \quad (5.22)$$

Example 5.46. Let us consider again Example 5.45 (box containing two white and three brown marbles). The probability of drawing two brown marbles is $P(AB) = P(A)P(B) = (3/5)(3/5) = 9/25$ or $P(AB) = P(A)P(B|A) = (3/5)(2/4) = 3/10$ according to whether the first marble is repositioned within the box or not.

5.7 Combinatorial Calculus

Calculating the probability of complex events from the probabilities of simple events can be facilitated by some techniques of combinatorial calculus, that are briefly recalled in this section.

Dispositions

The number of different ways of choosing k objects out of a set of n objects is called the number of *dispositions* of n objects of class k, and is indicated by $D_{n,k}$.

To understand how dispositions $D_{n,k}$ are calculated, it is convenient to begin by a simple example. In how many ways is it possible to draw $k = 2$ objects out of a set of $n = 3$ objects, $A, B,$ and C? One can easily verify that there are six ways: AB, AC, BC, BA, CA, CB, so that the dispositions are $D_{3,2} = 6$.

The general rule for calculating $D_{n,k}$ can be found according to the following considerations. Given a set of n objects, there are

n distinct ways of choosing the 1st object
$(n$-$1)$ distinct ways of choosing the 2nd object
...
$(n$-k+$1)$ distinct ways of choosing the kth object

As a consequence,

$$D_{n,k} = n \times (n-1) \times (n-2) \times \cdots \times (n-k+1). \quad (5.23)$$

Utilizing the factorial notation
$$n! = n \times (n-1) \times (n-2) \times \cdots \times 4 \times 3 \times 2 \times 1, \qquad 0! = 1,$$
and multiplying the right-hand side of (5.23) by $(n-k)!/(n-k)! = 1$, one gets the more synthetic expression
$$D_{n,k} = \frac{n!}{(n-k)!}. \qquad (5.24)$$

It is worth noting that two groups of objects that differ solely in the order of elements, such as AB and BA, are counted as two different dispositions.

Permutations

The *permutations* P_n of n objects are the different ways of ordering them. Permutations are a particular case of dispositions:
$$P_n = D_{n,n} = n! \qquad (5.25)$$

Example 5.47. Seven people can be distributed on seven available seats in $P_7 = 7! = 5040$ different ways.

Example 5.48. During laboratory sessions, a class of 30 students is divided into ten groups of three students. How many distinct ways of distributing the students into the ten groups are possible? The permutations of the 30 students are $P_{30} = 30!$ For each group, however, $3!$ permutations are equivalent. The total number of ways of forming different groups is thus
$$\frac{30!}{3!\,3!\,\cdots\,3!} = \frac{30!}{(3!)^{10}} \simeq 4 \times 10^{24}.$$

Combinations

The number of different ways of choosing k objects out of the set of n objects, without distinguishing the groups of k objects that only differ in their order (e.g., $AB \equiv BA$), is called the number of *combinations* of n objects of class k, and is indicated by $C_{n,k}$.

The combinations $C_{n,k}$ are calculated by dividing the dispositions $D_{n,k}$ (5.24) by the permutations P_k (5.25):
$$C_{n,k} = \frac{D_{n,k}}{P_k} = \frac{n!}{(n-k)!\,k!} = \binom{n}{k}. \qquad (5.26)$$

The difference between dispositions and combinations is particularly evident for $k = n$, where the dispositions are $D_{n,n} = n!$, whereas the combinations are $C_{n,n} = 1$.

5 Basic Probability Concepts

Example 5.49. Let us again consider the drawing, without reposition, of two marbles from a box containing two white and three brown marbles. The probability of the event $C =$ "both marbles are brown" has been previously calculated directly from the sample space, using the classical probability (Example 5.26), as well as by decomposing the event C into the product of two simpler events and using the rule for the probability of the product (Example 5.46). We calculate now the probability of the event C according to the classical probability, without building up the sample space, but using the techniques of combinatorial calculus:

$$\mathcal{P}(C) = \frac{\text{number of favorable outcomes}}{\text{total number of outcomes}} = \frac{\binom{3}{2}}{\binom{5}{2}} = \frac{3}{10}.$$

Newton Binomial

An important application of combinatorial techniques is the expression of the nth power of a binomial, $(a+b)^n$:

$$(a+b)^n = \underbrace{(a+b)\,(a+b)\,\cdots\,(a+b)}_{n \text{ factors}}$$

$$= a^n + \binom{n}{1}a^{n-1}b + \cdots + \binom{n}{n-1}ab^{n-1} + b^n$$

$$= \sum_{k=0}^{n} \binom{n}{k} a^{n-k}b^k. \qquad (5.27)$$

To understand the last equality, remember that

$$\binom{n}{k} = \binom{n}{n-k} \quad \text{and} \quad \binom{1}{0} = \binom{1}{1} = 1.$$

With reference to (5.27), the expressions $\binom{n}{k}$ are called *binomial coefficients*.

Problems

5.1. Two dice are contemporarily tossed. Build up the sample space and verify that the probability of obtaining "7" or "5" as the sum of the two numbers is $\mathcal{P}(7) = 1/6$ and $\mathcal{P}(5) = 1/9$, respectively.

5.2. Two cards are chosen from a deck of 40 cards. Calculate the total number of possible outcomes. Verify that the probability of obtaining two aces is $\mathcal{P} = 1/130$. Verify that the probability that only one card is an ace is $\mathcal{P} = 12/65$.

5.3. A die is tossed twice. Show that the probability of the event $B =$ "the number 4 appears at least once" is $P(B) = 11/36$.

Hints: B is the sum of two events: $A_1 =$ "4 appears at the first trial"; $A_2 =$ "4 appears at the second trial". Are the two events mutually exclusive?

5.4. A die is tossed five times. Show that the probability of the event $A =$ "the number 6 appears three times" is $P(A) = 125/3888$.

Hints: Using combinatorial techniques, calculate the number of different equivalent ways of obtaining "6" three times out of five tosses. Then decompose the event A into sums and products of simpler events.

5.5. A box contains ten white and five brown marbles. Four marbles are extracted by chance, without reposition. Show that the probability of the event $A=$"at least one marble is brown" is $P(A) = 11/13$.
Compare two different procedures.

1. A is the sum of four mutually exclusive events, $A = A_1 + A_2 + A_3 + A_4$, where $A_i=$ "i brown marbles are extracted". Suitably decompose each of the A_i events into sums and products of simpler events.
2. Consider the contrary event $\overline{A} =$ "no brown marbles are extracted", express it as a product $\overline{A} = B_1 B_2 B_3 B_4$, where $B_i =$ "the ith marble is white", and calculate its probability using the rules for the product of nonindependent events.

5.6. A coin is tossed six times. Show that the probability of the event $A =$ "the number of heads is larger than the number of tails" is $P(A) = 11/32$.
Compare two different procedures.

1. Express the event A in terms of sums and products of simpler events.
2. Exploit the symmetry properties of the problem, by comparing the probabilities of the three mutually exclusive events, $A =$ "more heads than tails", $B =$ "three heads and three tails", $C =$ "more tails than heads", and directly evaluating the probability of B.

6 Distributions of Random Variables

In Chap. 5, the basic concepts of probability theory have been presented in terms of events and probabilities of events. An alternative approach is based on random variables and distributions of random variables. Some probability distributions have already been phenomenologically introduced in Chap. 4. In this chapter, the general properties of the distributions of random variables are explored, and some of the distributions that are particularly relevant in data analysis procedures are studied in detail.

6.1 Binomial Distribution

The binomial distribution plays a central role in probability theory. It is used here to introduce the general topic of random variables and their distributions in a heuristic way.

Repeated Independent Trials

Let us consider an experiment where each trial can have only two possible outcomes, positive or negative, and let

p the probability of positive outcome.
$q = 1 - p$ the probability of negative outcome.

Let us now suppose that the experiment is repeated many times, and the outcome of each repetition is independent of the previous outcomes. Our goal is to evaluate the probability that, out of n experiments, k have a positive outcome (and $n - k$ a negative outcome). To this aim, let us first consider a simple example.

Example 6.1. Three bullets ($n = 3$) are shot at a target. The probability of hitting the target is p at each shot. What is the probability of hitting the target two times ($k = 2$)? Let us introduce the notation:

Event B = two shots out of three hit the target.
Event A_i = the ith shot hits the target.
Event $\overline{A}_i$ = the ith shot does not hit the target.

The event B can be decomposed into sums and products, $B = A_1 A_2 \overline{A}_3 + A_1 \overline{A}_2 A_3 + \overline{A}_1 A_2 A_3$, and, because the events A_i are independent, $\mathcal{P}(B) = pp(1-p) + p(1-p)p + (1-p)pp = 3p^2(1-p)$.

The problem of repeated trials can be generalized as follows. The event $B_k=$ "k outcomes out of n are positive" is the sum of $\binom{n}{k}$ combinations of mutually exclusive events. Each one of the $\binom{n}{k}$ terms is in turn the product of n independent events: k with positive outcome, and $n-k$ with negative outcome. Making use of (5.12) for the probability of the sum of mutually exclusive events, and of (5.22) for the probability of the product of independent events, one finds the probability of B_k:

$$\mathcal{P}(B_k) = \binom{n}{k} p^k q^{n-k} . \tag{6.1}$$

Example 6.2. A die is drawn $n = 7$ times. What is the probability of the event $B_2 = $ "the number 3 appears two times"? The probability of the outcome "3" for a die is $p = 1/6$. According to (6.1),

$$\mathcal{P}(B_2) = \binom{7}{2} \left(\frac{1}{6}\right)^2 \left(\frac{5}{6}\right)^5 = 0.234 .$$

The Binomial Distribution

In an experiment consisting of n independent trials, each one with probability p of positive outcome, the number k of positive outcomes can be considered as an integer variable, assuming the values $0 \leq k \leq n$. The probability that k out of n outcomes are positive is thus a function of the integer variable k:

$$\mathcal{P}_{np}(k) = \binom{n}{k} p^k q^{n-k} . \tag{6.2}$$

The probability function (6.2) is called a *binomial distribution*, because (6.2) is a term of the Newton binomial expansion (5.27), or *Bernoulli distribution*, after the name of the Swiss mathematician J. Bernoulli (1654–1705).

The binomial distribution depends on two parameters, n and p, that in (6.2) appear as indexes of $\mathcal{P}$. To simplify the notations, the indexes n and p, when not strictly necessary, are omitted in the following.

By summing the terms of the binomial distribution for all values of k from 0 to n, one gets

$$\sum_{k=0}^{n} \mathcal{P}(k) = \sum_{k=0}^{n} \binom{n}{k} p^k q^{n-k} = (p+q)^n = 1^n = 1 . \tag{6.3}$$

Actually, the sum of the probabilities of all the possible events is the probability of the certain event, say 1. This *normalization condition* holds for all probability distributions.

6.1 Binomial Distribution

When $p = 0.5$, also $q = 0.5$, and the binomial distribution becomes $\mathcal{P}(k) = \binom{n}{k} 0.5^n$, and is symmetrical with respect to $k = n/2$.

Example 6.3. The binomial distributions for $n = 6$ repeated trials and different values of the trial probability p are shown in Table 6.1 and in Fig. 6.1.

Table 6.1. Binomial distributions for $n = 6$ and different values of p (Example 6.3). Each line corresponds to one of the possible events.

	$p = 0$	$p = 0.1$	$p = 0.3$	$p = 0.5$	$p = 0.9$
$\mathcal{P}(k = 0)$	1	0.531	0.118	0.015	1×10^{-6}
$\mathcal{P}(k = 1)$	0	0.354	0.302	0.094	5×10^{-5}
$\mathcal{P}(k = 2)$	0	0.098	0.324	0.234	0.001
$\mathcal{P}(k = 3)$	0	0.015	0.185	0.312	0.015
$\mathcal{P}(k = 4)$	0	0.001	0.059	0.234	0.098
$\mathcal{P}(k = 5)$	0	5×10^{-5}	0.010	0.094	0.354
$\mathcal{P}(k = 6)$	0	1×10^{-6}	7×10^{-4}	0.015	0.531

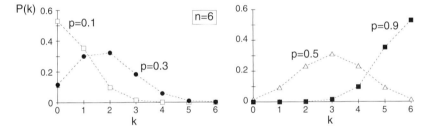

Fig. 6.1. Binomial distributions for $n = 6$ and different values of p (Example 6.3). The function $\mathcal{P}(k)$ is defined for integer values of k: the dashed lines are only a guide to the eye. For $p = 0.5$, the distribution is symmetrical with respect to $k = np = 3$.

Example 6.4. When a coin is tossed, the probability of "head" is $p = 0.5$. We want to evaluate the probability that, when the coin is tossed n times, one obtains exactly $n/2$ "heads". The binomial distribution giving the probability of k "heads" out of n tosses is shown in Fig. 6.2 for different values of n. The distributions of Fig. 6.2 (left) have the maximum for $k = n/2$; however, when n increases, the probability of obtaining exactly $k = n/2$ "heads" progressively decreases. This fact, apparently surprising, can be better understood by considering the distribution as a function of the frequency k/n, instead of k (Fig. 6.2, right). When n increases, the values k/n become more and more densely packed, and the values corresponding to a nonnegligible probability group around the central value $k/n = p = 0.5$. Otherwise stated, the fluctuations of the frequency k/n with respect to the central value $k/n = p = 0.5$ become progressively smaller.

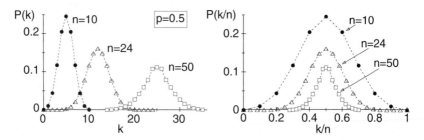

Fig. 6.2. Illustrating Example 6.4. Left: binomial distributions with probability $p = 0.5$ for different values of n. Right: the same distributions plotted as a function of the normalized frequency k/n. (The dashed lines are only a guide to the eye.)

Example 6.5. Let us consider two segments, the first one, AD, of length ℓ, and the second one, BC, of length $s = \ell/5$, contained within the first one. The probability that a point, chosen by chance within the segment AD, is contained in BC, is $p = s/\ell = 0.2$. Let us now consider n points chosen by chance in AD; the probability that k out of n points are contained in BC is given by the binomial distribution

$$P(k) = \binom{n}{k} p^k q^{n-k} .$$

Some binomial distributions for $p = 0.2$ are shown in Fig. 6.3. The distributions are asymmetrical, but the asymmetry decreases when n increases. The problem is formally identical if AD and BC are time intervals instead of segments, and is further considered when studying the Poisson distribution (Sect. 6.4).

The Limiting Case $n = 1$

A particular case of binomial distribution is encountered when $n = 1$, say when only one trial is attempted. The possible outcomes are two, positive or negative. Correspondingly, the random variable can only assume the values $k_1 = 0$ or $k_2 = 1$. The binomial distribution is:

$$\mathcal{P}(0) = 1 - p = q , \qquad \mathcal{P}(1) = p . \tag{6.4}$$

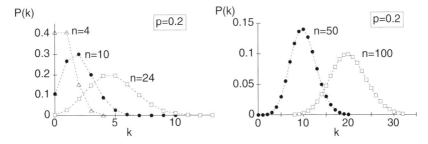

Fig. 6.3. Illustrating Example 6.5. Binomial distributions for $p = 0.2$ and different values of n. The asymmetry decreases when n increases. (The dashed lines are only a guide to the eye.)

Binomial Distribution for Histogram Columns

An interesting application of the binomial distribution concerns the height of the histogram columns. Let us suppose that the measurement of a quantity X has been repeated N times, and the corresponding histogram has been drawn (Sect. 4.3). Consider now the jth column of the histogram, where n_j^* measurements have been recorded (Fig. 6.4). If the N measurements were repeated, different histograms with different n_j^* values would be obtained; n_j^* is then a random variable. The histogram is a sample of the limiting distribution of the X variable. The probability p_j that a measure x is within the jth interval of the histogram can be calculated from the limiting distribution. The probability that n_j^* values are found within the jth interval is given by the binomial distribution:

$$\mathcal{P}_{Np_j}(n_j^*) = \binom{N}{n_j^*} p_j^{n_j^*} q_j^{N-n_j^*} . \tag{6.5}$$

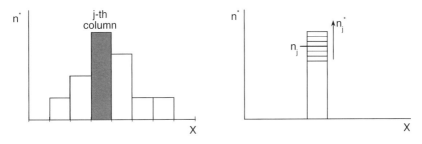

Fig. 6.4. Left: histogram for N repeated measurements of a quantity X. Right: the height n_j^* of the jth column is a random variable that follows the binomial distribution.

6.2 Random Variables and Distribution Laws

The present section is dedicated to a generalization of the concepts introduced in the previous Sect. 6.1. A *random variable* (r.v.) is a quantity that, in a given experiment, can assume an unpredictable value, within a set of possible values.

Example 6.6. Let us consider again the problem of repeated trials that led to the binomial distribution in Sect. 6.1. If a trial is repeated n times, $\mathcal{N} = n+1$ events have to be considered, corresponding to $k = 0, 1, 2, \ldots, n$ positive outcomes, respectively. The $n + 1$ events can be represented by a random variable K, that can assume all the $\mathcal{N}$ integer values from 0 to n.

A *discrete random variable* can assume a finite or a countably infinite number of values. For example, the number of "heads" obtained when a coin is tossed ten times is a discrete random variable that can assume only a finite number of values; the number of times a coin has to be tossed before "head" is obtained for the first time, is again a discrete random variable that can, however, assume a countably infinite number of values.

A *continuous random variable* can assume an uncountably infinite number of values. The lifetime of an electronic device is an example of a continuous random variable.

In the following, attention is mainly focused on the random phenomena that can be described by one random variable. In Sect. 6.8, random phenomena described by two or more random variables are briefly considered.

Distributions of Discrete Random Variables

A discrete random variable K can assume a finite or a countably infinite number of values $k_1, k_2, \ldots, k_j, \ldots$

A *distribution law* of a discrete random variable gives a value of probability for each value of the random variable:

$$\mathcal{P}(k_j) = p_j . \tag{6.6}$$

For a complete system of mutually exclusive events, the *normalization condition* holds:

$$\sum_j p_j = 1 . \tag{6.7}$$

The binomial distribution $\mathcal{P}_{np}(k)$ previously introduced in Sect. 6.1 is an example of a distribution of discrete random variable that can assume a finite number $\mathcal{N} = n + 1$ of possible values: $k = 0, 1, 2, \ldots, n$.

An example of a discrete random variable that can assume a countably infinite number of values is given in Problem 6.3. Another important case, the Poisson distribution, is considered in Sect. 6.4.

Distributions of Continuous Random Variables

A continuous random variable X can assume an uncountably infinite number of possible values x. One cannot express the distribution law of a continuous random variable in terms of a simple correspondence law such as (6.6). Actually, it is shown below that, for a single value of a continuous variable, it is always $\mathcal{P}(x) = 0$.

To express a distribution of a continuous random variable, two mathematical instruments can be used:

1. The *cumulative distribution function* (or simply *distribution function*) $F(x)$
2. The *probability density* $f(x)$, corresponding to the first derivative of the distribution function: $f(x) = \mathrm{d}F(x)/\mathrm{d}x$

The *cumulative distribution function* $F(x)$ gives the probability that the random variable X has a value not larger than x (Fig. 6.5, left):

$$F(x) = \mathcal{P}(X \leq x) . \tag{6.8}$$

The cumulative distribution function has the following properties:

(a) $F(x)$ is a nondecreasing function of x: $x_1 < x_2 \Rightarrow F(x_1) \leq F(x_2)$.
(b) When $x \to -\infty$, $F(x) \to 0$.
(c) When $x \to +\infty$, $F(x) \to 1$.

The cumulative distribution function can also be defined for a discrete random variable K (Fig. 6.5, right):

$$F(k) = \mathcal{P}(K \leq k) = \sum_{k_i \leq k} \mathcal{P}(k_i) . \tag{6.9}$$

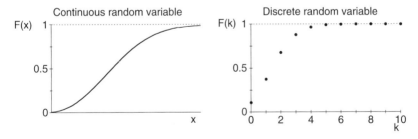

Fig. 6.5. Examples of cumulative distribution functions. Left, for a continuous r.v. X. Right, for the discrete r.v. K of a binomial distribution with $n = 10$ and $p = 0.2$.

If the cumulative distribution function $F(x)$ of a continuous random variable X is known, then the probability that X is included between two values x_1 and x_2 is (Fig. 6.6, left):

$$\mathcal{P}(x_1 < X \leq x_2) = \mathcal{P}(X \leq x_2) - \mathcal{P}(X \leq x_1) = F(x_2) - F(x_1). \quad (6.10)$$

According to (6.10), by progressively reducing the width of the interval $x_2 - x_1$, one finds

$$\mathcal{P}(x_1) = \lim_{x_2 \to x_1} [F(x_2) - F(x_1)] = 0. \quad (6.11)$$

For a continuous random variable X, the probability of a given value x_1 is thus always zero. This fact does not mean that the value x_1 cannot be obtained in a real experiment; it means instead that, when the number of repeated experiments increases, the relative frequency of the value x_1 progressively reduces to zero.

Because the probability of a single value x is always zero, it is equivalent, in (6.10), to write $\mathcal{P}(x_1 < X \leq x_2)$ or $\mathcal{P}(x_1 \leq X \leq x_2)$.

By definition, an impossible event has zero probability, $\mathcal{P} = 0$, and the certain event has unit probability, $\mathcal{P} = 1$. Notice, however, that an event with zero probability is not necessarily impossible, and an event with unit probability is not necessarily certain.

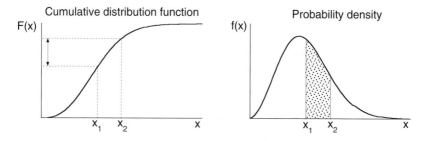

Fig. 6.6. Cumulative distribution function (left) and probability density (right) for a continuous random variable X.

The *probability density* $f(x)$ of a continuous random variable X is the first derivative of the cumulative distribution function $F(x)$:

$$f(x) = \lim_{\Delta x \to 0} \frac{F(x + \Delta x) - F(x)}{\Delta x} = \frac{\mathrm{d}F}{\mathrm{d}x}. \quad (6.12)$$

The probability that the random variable X has a value included between x_1 and x_2 is the integral of the probability density $f(x)$ calculated from x_1 to x_2 (Fig. 6.6, right):

$$\mathcal{P}(x_1 \leq X \leq x_2) = \int_{x_1}^{x_2} f(x')\,\mathrm{d}x' = F(x_2) - F(x_1). \quad (6.13)$$

Conversely, the cumulative distribution function $F(x)$ is obtained from the probability density by integration:

$$F(x) = \int_{-\infty}^{x} f(x')\,\mathrm{d}x' \;. \tag{6.14}$$

One can easily verify that:
(a) The probability density $f(x)$ is always nonnegative, $f(x) \geq 0$.
(b) The total area below the curve $f(x)$ is one, because it corresponds to the probability of the certain event (normalization condition):

$$\int_{-\infty}^{+\infty} f(x)\,\mathrm{d}x = 1 \;. \tag{6.15}$$

From the dimensional point of view, the probability density is the ratio between a probability, say a pure number, and the random variable. The probability density then has the inverse dimensions of the random variable.

Example 6.7. Uniform distribution. The uniform distribution (Fig. 6.7) is characterized by the following probability density.

$$f(x) = \begin{cases} 0 & \text{for } x < x_1 \;, \\ C & \text{for } x_1 \leq x < x_2 \;, \\ 0 & \text{for } x \geq x_2 \;, \end{cases} \tag{6.16}$$

where C is a constant whose value depends on the normalization condition:

$$\int_{-\infty}^{+\infty} f(x)\,\mathrm{d}x = 1 \quad \Rightarrow \quad C = \frac{1}{x_2 - x_1} \;. \tag{6.17}$$

By (6.14), one can easily verify that the cumulative distribution function is

$$F(x) = \begin{cases} 0 & \text{for } x < x_1 \;, \\ \dfrac{x - x_1}{x_2 - x_1} & \text{for } x_1 < x < x_2 \;, \\ 1 & \text{for } x \geq x_2 \;. \end{cases} \tag{6.18}$$

A uniform distribution has been used in Sect. 4.2 to represent the results of a single measurement of a physical quantity. The width of the interval $x_2 - x_1$ corresponds to the resolution Δx of the measurement.

Example 6.8. The normal distribution introduced in Sect. 4.3,

$$f(x) = \frac{1}{\sigma\sqrt{2\pi}} \exp\left[-\frac{(x-m)^2}{2\sigma^2}\right] \;, \tag{6.19}$$

is a probability density for the continuous random variable X. The properties of the normal distribution are studied in Sect. 6.5.

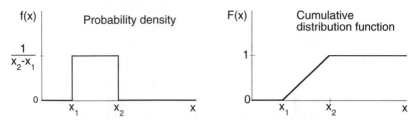

Fig. 6.7. Uniform distribution: probability density (left) and cumulative distribution function (right).

6.3 Numerical Characteristics of Distributions

A distribution law, such as a cumulative distribution function $F(x)$ or a probability density $f(x)$, completely characterizes the behavior of a random variable. In many cases, however, complete knowledge of the distribution is not required, and a few numerical parameters describing its main properties are sufficient.

Position Parameters

Several parameters can be used to characterize the position of the distribution along the axis representing the random variable. The most important position parameter is the *mean* (or *average value* or *expected value*).

For a discrete random variable K, the mean is defined as

$$m_k = \langle k \rangle = \sum_j k_j p_j , \qquad (6.20)$$

where the index j spans the possible values of K. Eq. (6.20) is a weighted average (Sect. 4.4): the values k_j of the discrete random variable are weighted by their probabilities p_j. The sum over the weights, which appears at the denominator of (4.39), is omitted here, because $\sum p_j = 1$.

The definition of the mean (6.20) is formally similar to the expression of the sample average of N experimental values (4.10), introduced in Sect. 4.3:

$$m_k^* = \frac{1}{N} \sum_{j=1}^{\mathcal{N}} k_j n_j = \sum_{j=1}^{\mathcal{N}} k_j p_j^* , \qquad (6.21)$$

where index j spans the $\mathcal{N}$ histogram columns and p_j^* is the statistical frequency of the value k_j. The connection between (6.21) and (6.20) is based on the concept of statistical probability (Sect. 5.3):

$$\lim_{N \to \infty} p_j^* = p_j , \quad \text{whence} \quad \lim_{N \to \infty} m_k^* = m_k .$$

For $N \to \infty$, the sample mean m_k^* tends to the mean m_k.

6.3 Numerical Characteristics of Distributions

For a continuous random variable X, the mean is defined as

$$m_x = \langle x \rangle = \int_{-\infty}^{+\infty} x\,f(x)\,\mathrm{d}x , \qquad (6.22)$$

where $f(x)$ is the probability density.

Comparing (6.20) with (6.22), one can notice that the sum is substituted by an integral, the value k_j of the discrete random variable K is substituted by the value X of the continuous random variable, and the probability p_j is substituted by the differential probability $f(x)\,\mathrm{d}x$.

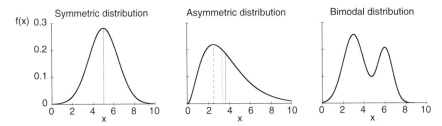

Fig. 6.8. Left: unimodal symmetric distribution; mean, mode and median coincide. Center: asymmetric distribution, right tailed; mode (dashed line), median (dotted line), and mean (continuous line) have different values. Right: bimodal distribution.

Two other position parameters are frequently used (Fig. 6.8).

The *mode* is the value of the random variable corresponding to the maximum value of the probability (for discrete random variables) or of the probability density (for continuous random variables). Distributions characterized by two or more maxima are called bimodal or manymodal, respectively.

The *median* is defined as follows. The probability that the random variable has a value smaller than the median is equal to the probability that the random variable has a value larger than the median.

Dispersion Parameters

In addition to position, another important characteristic of a distribution is its width. The parameters most frequently used to measure the width of a distribution are the *variance* D and its square root, the *standard deviation* $\sigma = \sqrt{D}$ (Fig. 6.9, left).

For a discrete random variable K, the variance is defined as

$$D_k = \langle (k - m_k)^2 \rangle = \sum_j (k_j - m_k)^2\, p_j , \qquad (6.23)$$

and the standard deviation is defined as

$$\sigma_k = \sqrt{D_k} = \sqrt{\sum_j (k_j - m_k)^2 \, p_j} \,. \tag{6.24}$$

As with the mean, also with the variance and standard deviation one can notice the formal similarity of (6.23) and (6.24) with (4.14) and (4.15), which define the sample variance D^* and the sample standard deviation σ^* of a histogram, respectively.

For a continuous random variable X, the variance is defined as

$$D_x = \left\langle (x - m_x)^2 \right\rangle = \int_{-\infty}^{+\infty} (x - m_x)^2 \, f(x) \, \mathrm{d}x \,, \tag{6.25}$$

and the standard deviation as

$$\sigma_x = \sqrt{D_x} = \sqrt{\int_{-\infty}^{+\infty} (x - m_x)^2 \, f(x) \, \mathrm{d}x} \,. \tag{6.26}$$

An alternative but equivalent expression of the variance, valid for both discrete and continuous random variables, can be obtained by the same procedure used for the sample variance in (4.16) (Sect. 4.3):

$$D_k = \langle k^2 \rangle - \langle k \rangle^2 \,, \qquad D_x = \langle x^2 \rangle - \langle x \rangle^2 \,. \tag{6.27}$$

It is worth noting that the standard deviation has the same dimensions as the random variable, whereas the variance has the dimensions of the square of the random variable.

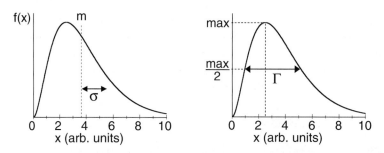

Fig. 6.9. Dispersion parameters: standard deviation σ (left) and full width at half maximum Γ (right). For the distribution in the figure, $\sigma = 1.96$, $\Gamma = 4.16$, $\gamma = \Gamma/2 = 2.08$ (arbitrary units).

Another parameter sometimes used to measure the width of a distribution is the *full-width at half maximum* (FWHM) Γ, say the width of the distribution measured in correspondence to half its maximum value (Fig. 6.9, right).

Alternatively, one sometimes quotes the *half-width at half maximum* $\gamma = \Gamma/2$.

The FWHM is typically used when the variance of a distribution cannot be defined, because the integral (6.25) is not convergent. An important example is the Cauchy–Lorentz distribution, that is considered in Sect. 6.7.

Moments of a Distribution

By generalizing the definitions of mean and variance, one can introduce two families of numerical parameters, the *initial moments* and the *central moments*. The knowledge of one of the two families completely characterizes a distribution (position, width, and shape).

The *initial moment of order s*, α_s, is the mean value of the sth power of the random variable. For discrete random variables,

$$\alpha_s = \langle k^s \rangle = \sum_j k_j^s p_j \,. \qquad (6.28)$$

For continuous random variables,

$$\alpha_s = \langle x^s \rangle = \int_{-\infty}^{+\infty} x^s \, f(x) \, dx \,. \qquad (6.29)$$

The mean is the initial moment of order one: $m_k = \alpha_1(K)$; $m_x = \alpha_1(X)$.

A general method for calculating the initial moments of distributions is presented in Appendix D.2.

The family of *central moments* is built up from the deviation s of the random variable with respect to its mean:

$$s_j = k_j - m_k \quad \text{or} \quad s_x = x - m_x \,.$$

The *central moment of order s*, μ_s, is the mean of the sth power of the deviation with respect to the mean. For discrete random variables,

$$\mu_s = \langle (k - m_k)^s \rangle = \sum_j (k_j - m_k)^s p_j \,. \qquad (6.30)$$

For continuous random variables,

$$\mu_s = \langle (x - m_x)^s \rangle = \int_{-\infty}^{+\infty} (x - m_x)^s \, f(x) \, dx \,. \qquad (6.31)$$

Central moments μ and initial moments α are connected by simple linear relations. For the low-order moments, the relations are given in (D.37) of Appendix D.2.

Notice that both initial and central moments of a given order s only exist if the integrals (6.29) and (6.31) for continuous variables, or the series

(6.28) and (6.29) for discrete variables with an infinite number of values, are convergent.

Let us now consider in some detail the meaning of the central moments of low order. For simplicity, only the case of discrete random variables is considered here, the extension to continuous random variables being straightforward (substitution of sums by integrals).

The *central moment of order 0* is always equal to 1 (normalization condition):

$$\mu_0 = \sum_j (k_j - m_k)^0 p_j = \sum_j p_j = 1 \;. \tag{6.32}$$

The *central moment of order 1* is always zero:

$$\mu_1 = \sum_j (k_j - m_k) p_j = \sum_j k_j p_j - m_k \sum_j p_j = 0 \;. \tag{6.33}$$

The *central moment of order 2* is the variance (6.23):

$$\mu_2 = D_k = \sum_j (k_j - m_k)^2 p_j \;. \tag{6.34}$$

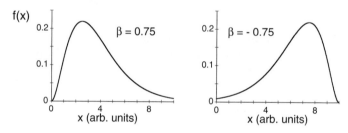

Fig. 6.10. Two asymmetric distributions with skewness $\beta = \mu_3/\sigma^3$ positive (left) and negative (right), respectively.

The *central moment of order 3*,

$$\mu_3 = \sum_j (k_j - m_k)^3 p_j \;, \tag{6.35}$$

as with all central moments of odd order, is zero for distributions symmetrical with respect to the mean, and can then be used to measure the asymmetry. The standard measure of asymmetry is the dimensionless *skewness coefficient*, defined as

$$\beta = \mu_3/\sigma^3 \;. \tag{6.36}$$

The coefficient β is positive or negative according to whether the tail of the distribution points to the right or to the left (Fig. 6.10).

The *central moment of order 4*,

6.3 Numerical Characteristics of Distributions

$$\mu_4 = \sum_j (k_j - m_k)^4 p_j , \quad (6.37)$$

is used to characterize the "peakedness" of the distribution, taking as reference the normal distribution, for which the dimensionless ratio μ_4/σ^4 has the value 3. The *kurtosis coefficient*, defined as

$$\gamma_2 = \mu_4/\sigma^4 - 3 , \quad (6.38)$$

is positive or negative according to whether the distribution is more or less peaked (less or more flat) than a normal distribution, respectively (Fig. 6.11).

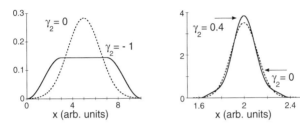

Fig. 6.11. Two symmetric distributions with kurtosis $\gamma_2 = \mu_4/\sigma^4 - 3$ negative (left) and positive (right). The dashed line is a normal distribution, for which $\gamma_2 = 0$.

Examples

Let us now evaluate the numerical characteristics of two distributions previously introduced, the binomial distribution and the uniform distribution.

Example 6.9. For the binomial distribution (Sect. 6.1)

$$\mathcal{P}(k) = \binom{n}{k} p^k q^{n-k} ,$$

one can demonstrate (Appendix D.3) that mean and variance are, respectively,

$$m = \sum_{k=0}^{n} k \, \frac{n!}{(n-k)!k!} p^k q^{n-k} = np , \quad (6.39)$$

$$D = \sum_{k=0}^{n} (k - m_k)^2 \, \frac{n!}{(n-k)!k!} p^k q^{n-k} = npq . \quad (6.40)$$

For a given value of p, the mean m increases proportionally to the number n of trials (see Figs. 6.1 and 6.2). The standard deviation $\sigma = \sqrt{D} = \sqrt{npq}$ increases proportionally to the square root of n. The relative width σ/m decreases proportionally to $1/\sqrt{n}$.

One can also demonstrate (Appendix D.3) that the central moments of order three and four are

$$\mu_3 = npq(q-p), \qquad \mu_4 = npq(1 + 3npq - 6pq), \tag{6.41}$$

respectively, so that the skewness and kurtosis coefficients are:

$$\beta = \frac{\mu_3}{\sigma^3} = \frac{q-p}{\sqrt{npq}}, \qquad \gamma_2 = \frac{\mu_4}{\sigma^4} - 3 = \frac{1 - 6pq}{npq}. \tag{6.42}$$

For $p = q = 0.5$, $\beta = 0$, because the distribution is symmetric. For $p < q$, $\beta > 0$, the tail of the distribution points to the right.

According to (6.42), when $n \to \infty$, both skewness and kurtosis coefficients tend to zero. Actually, it is shown in Sect. 6.6 that when $n \to \infty$ the shape of the binomial distribution tends to the shape of the normal distribution, for which $\beta = \gamma_2 = 0$.

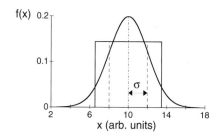

Fig. 6.12. Comparison between a uniform distribution and a normal distribution with the same mean $m = 10$ and standard deviation $\sigma = 2$. The area between $x = m - \sigma$ and $x = m + \sigma$ is 0.58 and 0.68 for the uniform distribution and the normal distribution, respectively. Both distributions are symmetric, with $\beta = 0$. The flatness parameter $\gamma_2 = \mu_4/\sigma^4 - 3$ is zero for the normal distribution, and $144/80 - 3$, say negative, for the uniform distribution.

Example 6.10. For the uniform distribution (6.16)

$$f(x) = \begin{cases} 0 & \text{for } x < x_1, \\ C & \text{for } x_1 \le x < x_2, \\ 0 & \text{for } x \ge x_2, \end{cases}$$

the constant C, determined by the normalization condition (6.17), is $C = 1/\Delta x$, where $\Delta x = 1/(x_2 - x_1)$. The mean m is

$$m = \int_{-\infty}^{+\infty} x f(x)\, dx = \frac{1}{\Delta x} \int_{x_1}^{x_2} x\, dx = \frac{x_1 + x_2}{2}. \tag{6.43}$$

By a similar procedure (Appendix D.4), one can calculate the variance and the other central moments:

$$\mu_2 = D = \frac{(\Delta x)^2}{12}, \quad \mu_3 = 0, \quad \mu_4 = \frac{(\Delta x)^4}{80}. \tag{6.44}$$

The area included between $x = m - \sigma$ and $x = m + \sigma$ is 0.58 (to be compared with the value 0.68 of the normal distribution, Fig. 6.12). The full-width at half maximum is $\Gamma = \Delta x$. The half-width at half maximum $\gamma = \Gamma/2$ is connected to the standard deviation σ by the relation: $\sigma = \gamma/\sqrt{3}$.

The uniform distribution has been used in Chap. 4 to describe the uncertainty due to the resolution of a single measurement (Sect. 4.2) or due to the estimation of systematic errors (Sect. 4.5). In both cases, the uncertainty δX is expressed, by convention, as the standard deviation of the uniform distribution, $\delta X = \Delta x / \sqrt{12}$.

6.4 Poisson Distribution

An important distribution of a discrete random variable is the *Poisson distribution* (after the name of the French mathematician S. Poisson, 1781–1840):

$$\mathcal{P}_a(k) = \frac{a^k}{k!} e^{-a}. \tag{6.45}$$

The Poisson distribution is sometimes used to approximate the binomial distribution, but its most relevant applications in physics concern the counting of some kinds of random phenomena, such as radioactive decays.

Before examining the applications of the Poisson distribution, it is convenient to study its mathematical properties. The random variable can assume any nonnegative integer value, $k \geq 0$. The set of k values is then countably infinite. The Poisson distribution depends on only one positive parameter a. The normalization condition is always satisfied by the Poisson distribution, because

$$\sum_{k=0}^{\infty} \frac{a^k}{k!} e^{-a} = e^{-a} \sum_{k=0}^{\infty} \frac{a^k}{k!} = e^{-a} e^a = 1. \tag{6.46}$$

Some Poisson distributions for different values of the a parameter are shown in Fig. 6.13. As one can easily see, when the value of a increases, the position of the distribution shifts towards higher k values, and the distribution becomes progressively larger and less asymmetric. These properties are quantitatively described by the numerical characteristics of the distribution.

Numerical Characteristics of the Poisson Distribution

One can demonstrate (Appendix D.5) that the mean and variance of the Poisson distribution are:

$$m = a, \quad D = \mu_2 = a. \tag{6.47}$$

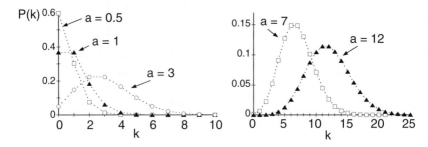

Fig. 6.13. Poisson distributions for different values of the a parameter. Notice that both horizontal and vertical scales of the two graphs are different. The dashed lines are only a guide to the eye.

The a parameter is thus equal to both the mean and the variance. This property characterizes the Poisson distribution, and represents a criterion often used to evaluate whether a statistical sample is compatible with a Poisson distribution. When a increases, the standard deviation $\sigma = \sqrt{D} = \sqrt{a}$ increases, and the relative width $\sigma/m = 1/\sqrt{a}$ decreases.

One can also demonstrate (Appendix D.5) that the skewness and kurtosis coefficients are

$$\beta = \frac{\mu_3}{\sigma^3} = \frac{1}{\sqrt{a}}, \quad \gamma_2 = \frac{\mu_4}{\sigma^4} - 3 = \frac{1}{a}. \tag{6.48}$$

When a increases, both coefficients decrease, and the shape of the Poisson distribution tends to the shape of the normal distribution.

Poisson Distribution and Binomial Distribution

It was shown in Sect. 6.3 that the mean of the binomial distribution is $m = np$. One can demonstrate that, when $n \to \infty$ and $p \to 0$, with the constraint that $m = np$ is finite, the binomial distribution tends to a Poisson distribution. To this aim, let it be from the beginning $a = m = np$, so that $p = a/n$ and $q = 1 - p = 1 - a/n$. The binomial distribution can be expressed as

$$\begin{aligned}\mathcal{P}(k) &= \frac{n(n-1)\cdots(n-k+1)}{k!}\left(\frac{a}{n}\right)^k \left(1 - \frac{a}{n}\right)^{n-k}\\ &= \frac{n(n-1)\cdots(n-k+1)}{n^k}\frac{a^k}{k!}\left(1-\frac{a}{n}\right)^n\left(1-\frac{a}{n}\right)^{-k}.\end{aligned} \tag{6.49}$$

The first and fourth factors in the last line of (6.49) tend to 1 when $n \to \infty$. As for the third factor,

$$\lim_{n\to\infty}\left(1-\frac{a}{n}\right)^n = e^{-a}, \tag{6.50}$$

so that, for $n \to \infty$ and $np =$ constant,

$$\mathcal{P}_{n,p}(k) \longrightarrow \mathcal{P}_a(k) \ . \tag{6.51}$$

In practice, the Poisson distribution is used to approximate the binomial distribution when contemporarily n is large (typically $n > 100$) and p is small (typically <0.05). The accuracy of the approximation can be evaluated by the examples of Fig. 6.14.

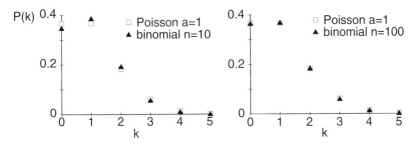

Fig. 6.14. Comparison between a Poisson distribution with $a = 1$ (squares) and two binomial distributions (triangles) of equal mean $m = np = 1$ and different values of n. Left: $n = 10$, $p = 0.1$. Right: $n = 100$, $p = 0.01$.

Example 6.11. In a factory, $n = 50$ similar machines are working. The probability that a machine breaks down during a shift is $p = 0.04$. The probability that $k = 4$ machines break down during a shift is exactly given by the binomial distribution,

$$\mathcal{P}_{50,0.04}(4) = \frac{50!}{46!\ 4!}\ 0.04^4\ 0.96^{46} = 0.09016\ ,$$

but can be approximated by the Poisson distribution, with $a = np = 2$:

$$\mathcal{P}_2(4) = \frac{2^4}{4!}\ e^{-2} = 0.0902\ .$$

Stationary Poisson Processes

Let us now resume and elaborate the application of the binomial distribution considered in Example 6.5 of Sect. 6.1. In a time (or space) interval of length ΔT, n independent events randomly occur. Two or more events cannot, however, occur at the same time. The probability that one of the n events, chosen by chance, occurs in a subinterval Δt is $p = \Delta t/\Delta T$. The probability that k events occur within the subinterval Δt is given by the binomial distribution

$$\mathcal{P}_{np}(k) = \binom{n}{k} \left[\frac{\Delta t}{\Delta T}\right]^k \left[1 - \frac{\Delta t}{\Delta T}\right]^{n-k} . \tag{6.52}$$

Let us now suppose that the duration of the interval ΔT and the number n of events proportionally increase, and introduce a density parameter $\lambda = n/\Delta T$ (average number of events per unit time interval). When ΔT increases, the probability $p = \Delta t/\Delta T$ proportionally decreases, but the average number of events within the subinterval Δt, say $m = np$, is fixed. In this case, as shown above, the binomial distribution can be replaced by a Poisson distribution.

It is, however, possible to completely give up knowledge of the parameters ΔT, n, and p, and uniquely refer to the density parameter $\lambda = n/\Delta T$. The probability that k events occur within the subinterval Δt is given by the Poisson distribution, with mean $a = np = \lambda \Delta t$:

$$\mathcal{P}_a(k) \;=\; \frac{a^k}{k!}\, e^{-a} \;-\; \frac{(\lambda \Delta t)^k}{k!}\, e^{-\lambda \Delta t}\,. \tag{6.53}$$

Many natural random phenomena are characterized by an average number of independent events per unit time (or space interval), but cannot be described by the binomial distribution (6.52) because one cannot attribute a value to the parameters n and p. These phenomena are called *Poisson processes*, and are described by the Poisson distribution (6.53), where $a = \lambda \Delta t$.

Example 6.12. The calls to a telephone exchange arrive randomly in time. The average number is $\lambda = 150$ calls per hour. The average number of calls within the time interval $\Delta t = 1$ minute is $a = \lambda \Delta t = 150(1/60) = 2.5$. The probability that $k = 6$ calls arrive within the interval $\Delta t = 1$ minute is given by the Poisson distribution:

$$\mathcal{P}(6) \;=\; \frac{2.5^6}{6!}\, e^{-2.5} \;=\; 0.278\,.$$

Counting Statistics

The Poisson distribution (6.53) is frequently utilized in physics to analyze the results of experiments based on the counting of random events (Sect. 1.4). Let us focus on the case of cosmic rays, very fast particles (such as protons or helium nuclei) that can traverse the atmosphere coming from space. Cosmic rays can be detected by suitable instruments, such as Geiger counters (Sect. 3.6).

The number of particles detected within the time Δt is a random variable k. Prolonged observations have, however, led to the conclusion that the average number λ of detected particles per unit time is a constant. The random variable k is thus governed by a Poisson distribution (6.53), with $a = \lambda \Delta t$. If λ were a priori known, one would also know $a = \lambda \Delta t$, and one could calculate the probability of a given number k of counts within the time interval Δt by (6.53).

In practice, one faces the reverse problem. One tries to determine the value of λ from counting experiments, and to evaluate its uncertainty. The situation is similar to that of Sect. 4.3, where, starting from a finite number of measurements of a physical quantity, we estimated the mean value of the limiting distribution and evaluated the uncertainty of the estimate. Here, the random variable is discrete instead of continuous, and the limiting distribution is Poisson instead of normal.

Single Measurements

Let us suppose that only one measurement is performed within the time interval Δt. The result is a number k of counts. The single measured value k corresponds to the sample mean of the random variable: $m_k^* = a^* = k$. As anticipated in Sect. 4.3 and demonstrated in Sect. 7.2, the best estimate of the limiting mean is the sample mean; in our case, the best estimate of a is $a^* = k$, say

$$\tilde{a} = k \text{ so that } \tilde{\lambda} = k/\Delta t \,. \tag{6.54}$$

The estimate $\tilde{a}$ has random character, and is thus affected by uncertainty. According to the conventions of Chap. 4, one assumes the standard deviation of the distribution of sample means as a measure of the uncertainty. In the present case, the sample mean coincides with the single measurement, $a^* = k$, and the limiting distribution of the sample means coincides with the limiting distribution of single counts, say the Poisson distribution (6.53). The Poisson distribution has the particular property that mean and variance coincide, so that $\sigma = \sqrt{a}$. The best estimate of the standard deviation is thus:

$$\tilde{\sigma} = \sqrt{\tilde{a}} = \sqrt{k} \,, \text{ so that } \delta\lambda = \sqrt{k}/\Delta t \,. \tag{6.55}$$

The uncertainty on the number of counts is equal to the square root of the number of counts. The relative uncertainty of the number of counts, and hence of the parameter λ, decreases inversely to the square root of the number of counts:

$$\delta\lambda/\tilde{\lambda} = 1/\sqrt{k} \,. \tag{6.56}$$

Example 6.13. The *radioactive decay* of an atom is a random process. If N nuclei of a radioactive isotope are present at time t, the average variation of the number of original isotopes in the time interval dt is $dN = -N\alpha\,dt$, where α is the *disintegration constant*. The average number of radioactive nuclei decreases according to the exponential law: $N(t) = N_0 \exp(-\alpha t)$. The average number λ of decays per unit time progressively decreases. If, however, the disintegration constant α is small, the parameter λ can be considered constant with good approximation for relatively long time intervals. In this case, if N_1 nuclei are present at the beginning of the experiment, the average number of decays in the time interval Δt, $a = \lambda \Delta t$, is given by

$$a = N_1\left[1 - e^{-\alpha \Delta t}\right] \simeq N_1\left[1 - 1 + \alpha \Delta t\right] = N_1 \alpha \Delta t \,, \tag{6.57}$$

where the exponential has been approximated by its expansion truncated at the second term: $\exp(-\alpha \Delta t) \simeq 1 - \alpha \Delta t$. The number k of decays in the time interval Δt is a random variable obeying the Poisson distribution (6.53), with $a = N_1 \alpha \Delta t$. If k events are counted in the interval Δt, then one can estimate $\tilde{a} = k \pm \sqrt{k}$.

It is worth noting that, in counting Poisson events, the uncertainty can be estimated from a single measurement, because mean and variance of the Poisson distribution have the same value. On the contrary, to estimate the uncertainty due to random fluctuations of the measure of a constant quantity (Sect. 4.3), it is necessary to measure at least two values, because mean and variance of the limiting distribution, typically normal, are uncorrelated.

Repeated Measurements

Let us now suppose that the counting is repeated N times, each time during a time interval Δt. N values k_i will be obtained, and one can calculate the sample mean and variance

$$m_k^* = \frac{1}{N} \sum_{i=1}^{N} k_i \ , \quad D_k^* = \frac{1}{N} \sum_{i=1}^{N} (k_i - m_k^*)^2 \ . \tag{6.58}$$

The degree of agreement between the values m_k^* and D_k^* is a first test to verify whether the limiting distribution can be assumed as a Poisson distribution. We again assume the sample mean m_k^* as the best estimate of the a parameter, and the standard deviation of the distribution of sample means as its uncertainty:

$$\tilde{a} = m_k^* \ , \quad \delta \tilde{a} = \sigma[m_k^*] = \frac{\sigma[k]}{\sqrt{N}} \simeq \frac{\sqrt{D_k^*}}{\sqrt{N}} \ . \tag{6.59}$$

When N increases, the uncertainty decreases. It is, however, completely equivalent to repeat the counting N times in a short time interval Δt, or to perform a unique counting during a long time interval $N \Delta t$.

Example 6.14. A Geiger counter records the impulses due to cosmic rays and natural radioactivity every $\Delta t = 15$ seconds.

In a first experiment, the counter works for about 1.5 hours, corresponding to $N = 361$ intervals Δt. The sample frequency of the count values k is shown by the vertical bars in Fig. 6.15, left. Sample mean and variance are $m_k^* = 4.629$ and $D_k^* = 4.931$, respectively. The best estimate (6.59) of the mean value is $\tilde{a} = 4.63 \pm 0.11$, corresponding to $\tilde{a}/\Delta t = 0.308 \pm 0.007$ counts per second; the corresponding Poisson distribution is shown as full circles in Fig. 6.15, left.

In a second experiment, the counter works for about eight hours, corresponding to $N = 28613$ intervals Δt. The sampling frequency of the count values k is shown by the vertical bars in Fig. 6.15, right. Sample mean and variance are $m_k^* = 4.592$ and $D_k^* = 4.638$, respectively. The best estimate (6.59)

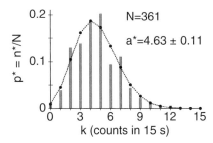 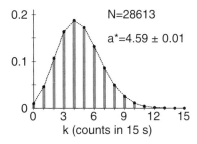

Fig. 6.15. Illustrating Example 6.14. Two different samplings of a Poisson distribution relative to the counts recorded in $\Delta t = 15$ s by a Geiger counter (vertical bars): sampling for 1.5 hours (left) and 8 hours (right). The full circles are the Poisson distribution estimated from the samples (the dashed lines are a guide to the eye).

of the mean value is $\tilde{a} = 4.59 \pm 0.01$, corresponding to $\tilde{a}/\Delta t = 0.306 \pm 0.0007$ counts per second; the corresponding Poisson distribution is shown as full circles in Fig. 6.15, right. By increasing the number N of time intervals Δt, the uncertainty of the estimated values of a has been reduced. The a values obtained in the two experiments are consistent.

Alternatively, if in both experiments a unique count had been done (one hour and a half long or eight hours long, respectively) the values $k = 1671$ and $k = 131385$ would have been obtained, respectively, corresponding to 0.308 ± 0.007 and 0.306 ± 0.0008 counts per second, in agreement with the previous procedure.

Poisson Distribution for Histogram Columns

It has been shown in Sect. 6.1 (Fig. 6.4) that, if a histogram is built up from a fixed number N of measurements, the distribution of the values n_j^* within a given column is binomial.

In some cases, however, instead of fixing the number N of measurements, one prefers to establish the time interval Δt during which the measurements are performed. In this case, the n_j^* values obey the Poisson distribution. The mean value is $a = n_j$, where n_j is the number of counts in the jth column, according to the limiting distribution of the quantity X: $n_j = p_j N$.

A generic experimental value n_j^* approximates the limiting value n_j with an uncertainty δn_j equal to the standard deviation of the Poisson distribution, $\sigma_j = \sqrt{a} = \sqrt{n_j}$, that can be estimated as $\tilde{\sigma}_j \simeq \sqrt{n_j^*}$.

6.5 Normal Distribution

The *normal distribution* (or *Gauss distribution*, after the name of the German mathematician C. F. Gauss, 1777–1855) is a distribution of a continuous random variable that plays a fundamental role in probability theory and in data analysis procedures. It has already been noted in Sect. 4.3 that the normal distribution is generally a good approximation of the limiting distribution of histograms relative to measurements of physical quantities.

The probability density of the normal distribution,

$$f(x) = \frac{1}{\sigma\sqrt{2\pi}} \exp\left[-\frac{(x-m)^2}{2\sigma^2}\right], \qquad (6.60)$$

depends on two parameters, m (real) and σ (real and positive). For any values of the parameters m and σ, the graph of the normal distribution is bell-shaped, and symmetric with respect to the value $x = m$ (Fig. 6.14).

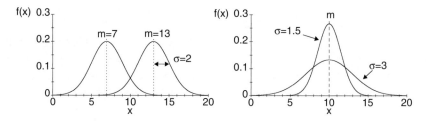

Fig. 6.16. Normal distributions: with the same value σ and different values m at the left, with the same value m and different values σ at the right.

Normalization Condition of the Normal Distribution

Let us first verify that the probability density (6.60) is normalized to one:

$$\frac{1}{\sigma\sqrt{2\pi}} \int_{-\infty}^{+\infty} \exp\left[-\frac{(x-m)^2}{2\sigma^2}\right] dx = 1. \qquad (6.61)$$

To this aim, by substituting

$$t = \frac{x-m}{\sigma\sqrt{2}}, \quad \text{whence } dx = \sigma\sqrt{2}\,dt, \qquad (6.62)$$

one transforms (6.61) into

$$\frac{1}{\sqrt{\pi}} \int_{-\infty}^{+\infty} e^{-t^2} dt = 1. \qquad (6.63)$$

The integral in (6.63) is the *Eulero–Poisson integral*, whose value is calculated in Appendix D.6:

$$\int_{-\infty}^{+\infty} e^{-t^2} dt = \sqrt{\pi}. \qquad (6.64)$$

Numerical Characteristics of the Normal Distribution

The numerical characteristics of the normal distribution can be easily calculated by means of substitution (6.62) and taking into account the Eulero–Poisson integral (6.64). The details of the calculations are given in Appendix D.6; here the main results are synthesized.

The mean of the distribution is equal to the parameter m (Fig. 6.16, left). For the central moments, the following recurrence relation holds:

$$\mu_s = (s-1)\sigma^2 \mu_{s-2} \qquad (s \geq 2) . \tag{6.65}$$

Because $\mu_1 = 0$, according to (6.65) all central moments of odd order are zero as well; this is not surprising, inasmuch as the normal distribution is symmetric with respect to $x = m$.

Because $\mu_0 = 1$ (normalization integral), from (6.65) one can see that the variance of the normal distribution is $D = \mu_2 = \sigma^2$, hence the parameter σ in (6.60) represents the standard deviation.

The maximum of the normal distribution corresponds to the value $x = m$. One can easily check that

$$f_{\max}(x) = f(m) = \frac{1}{\sigma\sqrt{2\pi}} = \frac{0.399}{\sigma} ; \tag{6.66}$$

when σ increases, the distribution becomes larger and lower (Fig. 6.16, right). At a distance σ from the mean m, the value of the distribution is:

$$f(m+\sigma) = f(m-\sigma) = \frac{1}{\sigma\sqrt{2\pi}} e^{-1/2} = 0.6\, f(m) . \tag{6.67}$$

Again from (6.65), one can see that $\mu_4 = 3\sigma^4$. The kurtosis coefficient $\gamma_2 = \mu_4/\sigma^4 - 3$ is zero.

Finally, it is easy to verify that the full-width at half maximum is $\Gamma = 2.35\,\sigma$, whence $\gamma = \Gamma/2 = 1.17\,\sigma$.

Calculating Probabilities for the Normal Distribution

In practical applications, it is required to calculate the probability that the random variable x is included between two given values α and β:

$$P(\alpha < x < \beta) = \frac{1}{\sigma\sqrt{2\pi}} \int_\alpha^\beta \exp\left[-\frac{(x-m)^2}{2\sigma^2}\right] dx . \tag{6.68}$$

The integral in (6.68) depends on the parameters m and σ, as well as on the values α and β. To evaluate the integral, let us first make the substitution

$$z = \frac{x-m}{\sigma} \quad \text{whence} \quad dx = \sigma\, dz . \tag{6.69}$$

Notice that the substitution (6.69) differs from the previous substitution (6.62) by a factor $\sqrt{2}$. The new variable z in (6.69) is the standard deviation of the old variable x with respect to the mean m, measured in units σ; it is called the *standard normal variable*. The standard normal variable is dimensionless, because it is defined in (6.69) as the ratio between two quantities with the same dimensions.

To determine the distribution $\phi(z)$ of the variable z, one can notice that the probability elements must be equal with respect to both variables, x and z, say $f(x)\,\mathrm{d}x = \phi(z)\,\mathrm{d}z$. The distribution of z, the *standard normal density*, is thus

$$\phi(z) = \frac{1}{\sqrt{2\pi}} \exp\left[-\frac{z^2}{2}\right]. \tag{6.70}$$

The standard normal density has zero mean, $m_z = 0$, and unit standard deviation, $\sigma_z = 1$ (Fig. 6.17, left).

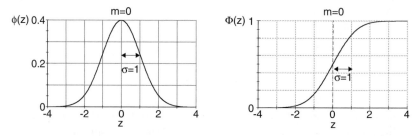

Fig. 6.17. Standard normal distribution: probability density $\phi(z)$ (left) and cumulative distribution function (right).

Once the limits of the integral have been substituted,

$$\alpha \to z_\alpha = \frac{\alpha - m}{\sigma}, \quad \beta \to z_\beta = \frac{\beta - m}{\sigma}, \tag{6.71}$$

the probabilities can be calculated as

$$P(\alpha < x < \beta) = \frac{1}{\sqrt{2\pi}} \int_{z_\alpha}^{z_\beta} \exp\left[-\frac{z^2}{2}\right] \mathrm{d}z. \tag{6.72}$$

The integral in (6.72) cannot be analytically calculated, so its values are tabulated in various ways.

(a) The *standard cumulative distribution* (Fig. 6.17, right)

$$\Phi(z) = \frac{1}{\sqrt{2\pi}} \int_{-\infty}^{z} \exp\left[-\frac{z'^2}{2}\right] \mathrm{d}z', \tag{6.73}$$

can be tabulated, so that

$$P(\alpha < x < \beta) = \Phi(z_\beta) - \Phi(z_\alpha). \tag{6.74}$$

(b) Alternatively, one can find the tables of the function

$$\Phi^*(z) = \frac{1}{\sqrt{2\pi}} \int_0^z \exp\left[-\frac{z'^2}{2}\right] dz' . \qquad (6.75)$$

(c) For measurement uncertainties, an interval symmetric with respect to the mean is considered, and one frequently makes use of the function

$$\tilde{\Phi}(z) = \frac{1}{\sqrt{2\pi}} \int_{-z}^z \exp\left[-\frac{z'^2}{2}\right] dz' . \qquad (6.76)$$

The meaning of the three functions $\Phi(z), \Phi^*(z)$, and $\tilde{\Phi}(z)$ is schematically illustrated in Fig. 6.18.

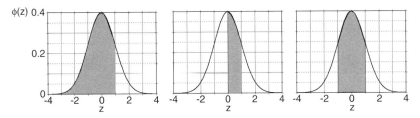

Fig. 6.18. Integrals of the standard normal distribution. The grey area represents the value of the three functions $\Phi(z)$ (left), $\Phi^*(z)$ (center), and $\tilde{\Phi}(z)$ (right) for $z = 1$.

By using the tables of the integrals of the standard normal density (Appendix C.3), one can easily find that, for any normal distribution:

$$\begin{aligned}
\mathcal{P}\,(m < x < m + \sigma) &= \mathcal{P}\,(0 < z < 1) = \Phi^*(1) = 0.3413\,, \\
\mathcal{P}\,(m < x < m + 2\sigma) &= \mathcal{P}\,(0 < z < 2) = \Phi^*(2) = 0.4772\,, \\
\mathcal{P}\,(m < x < m + 3\sigma) &= \mathcal{P}\,(0 < z < 3) = \Phi^*(3) = 0.4987\,.
\end{aligned}$$

The probabilities that the random variable x is included within an interval centered on m, of width 2σ, 4σ, and 6σ are, respectively (Fig. 6.19):

$$\begin{aligned}
\mathcal{P}\,(m - \sigma < x < m + \sigma) &= \mathcal{P}\,(-1 < z < 1) = \tilde{\Phi}(1) = 0.6826\,, \\
\mathcal{P}\,(m - 2\sigma < x < m + 2\sigma) &= \mathcal{P}\,(-2 < z < 2) = \tilde{\Phi}(2) = 0.9544\,, \\
\mathcal{P}\,(m - 3\sigma < x < m + 3\sigma) &= \mathcal{P}\,(-3 < z < 3) = \tilde{\Phi}(3) = 0.9974\,.
\end{aligned}$$

The integrals of the standard normal density are frequently referred to as the *error function*, erf(y). There is, however, no clear agreement on the definition of the error function: the y variable can correspond to the variable t of (6.62) or to the variable z of (6.69), and the integral extrema can be $(0, y)$ or $(-y, y)$.

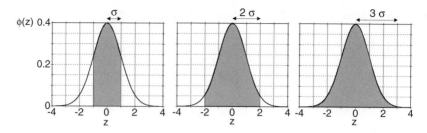

Fig. 6.19. The probability that $-1 < z < +1$ is 0.6826 (left); the probability that $-2 < z < +2$ is 0.9544 (center); the probability that $-3 < z < +3$ is 0.9974 (right).

6.6 Meaning of the Normal Distribution

The normal distribution plays a fundamental role in probability theory because it represents a limiting distribution in a number of relevant situations.

One can demonstrate that, when a random variable S can be decomposed into a sum of a large number of other random variables Y_i, the distribution of the variable S is in many cases normal. The conditions under which the sum S has a normal distribution are defined by a group of theorems, globally called the *central limit theorem*.

Central Limit Theorem

A version of the central limit theorem, particularly suited to data analysis applications, is given here.

Let us suppose that a random variable S can be expressed as a linear combination of n independent random variables Y_i,

$$S = \alpha_1 Y_1 + \alpha_2 Y_2 + \cdots + \alpha_n Y_n = \sum_{i=1}^{n} \alpha_i Y_i. \quad (6.77)$$

Let us further suppose that the means m_i and variances D_i of the variables Y_i exist and are of the same order of magnitude for the different variables Y_i, and also the coefficients α_i are of the same order of magnitude. One can then demonstrate (with some restrictions of limited practical relevance) that, when the number n of terms in the sum (6.77) increases, the distribution of the random variable S tends to a normal distribution, whose mean and variance are

$$m_S = \sum_i \alpha_i m_i, \qquad D_S = \sum_i \alpha_i^2 D_i, \quad (6.78)$$

respectively, independently of the shape of the distributions of the random variables Y_i.

6.6 Meaning of the Normal Distribution

Example 6.15. It was shown in Sect. 4.1 that the histograms of repeated measurements of a physical quantity tend to become bell-shaped when the number of measurements increases. Gauss invented the normal distribution to give a theoretical interpretation of this behavior. His argument can be recast as follows. The single measure x differs from the true value X_v by an error E: $x = X_v + E$. The error E is due to many random factors ϵ_i, independent and of the same extent, $E = \epsilon_1 + \epsilon_2 + \epsilon_3 + \cdots$, and can thus be considered as an example of the variable S of (6.77). The distribution of E is normal, as well as the distribution of $x = X_0 + E$. The possibility of decomposing the error E into the sum of a sufficiently large number of elementary contributions ϵ_i is generally not evident, and the Gauss argument has no general validity.

Example 6.16. Let us consider the sample mean m^* of a set of N measures of a physical quantity (Sect. 4.1):

$$m^* = \frac{1}{N} \sum_{i=1}^{N} x_i = \sum_{i=1}^{N} \frac{1}{N} x_i . \tag{6.79}$$

The mean m^* can be identified with the variable S of (6.77), with $\alpha_i = 1/N$. In this case, all terms x_i have the same distribution, that generally, but not necessarily, is normal. The central limit theorem states that the distribution of sample means m^* can be assumed to be normal, if the number N of measurements is large enough.

Example 6.17. An experiment consists of n independent repetitions of the same trial, according to the scheme leading to the binomial distribution of Sect. 6.1. A two-valued random variable Y is associated with each trial: $Y = 0$ if the trial has negative outcome, and $Y = 1$ if the trial has positive outcome. The binomial random variable K, corresponding to the number of positive outcomes out of n trials, can be expressed as a sum

$$K = \sum_{i=1}^{n} Y_i ,$$

and can be identified with the variable S of (6.77). One thus expects that the binomial distribution of the random variable K tends to a normal shape when n increases.

Discrete and Continuous Random Variables

In many applications of the central limit theorem, one deals with the sum of discrete random variables, whereas the limiting normal distribution is a distribution of a continuous random variable. The comparison of distributions of discrete and of continuous random variables was already considered in Sect. 4.3, when the relation between experimental histograms and limiting distributions was treated. The problem is studied here in a more systematic way.

The distributions of discrete and continuous random variables are intrinsically different. For a discrete random variable, the distribution law directly gives the probability $\mathcal{P}(k_j)$, whereas for a continuous random variable the distribution law is represented by a probability density $f(x)$. To compare a distribution of a discrete random variable with a distribution of a continuous random variable, different procedures can be used, which are listed below and illustrated in Fig. 6.20. To facilitate the comparison, the values of the discrete random variables are here indicated by x_j instead of k_j.

(a) The distribution of the discrete variable $\mathcal{P}(x_j)$ (Fig. 6.20, left) is substituted by an area-normalized histogram. With each value x_j of the discrete variable, a column of area equal to $\mathcal{P}(x_j)$ and base $\Delta x_j = (x_{j+1} - x_{j-1})/2$ is associated (Fig. 6.20, center). The histogram can now be directly compared with the probability density $f(x)$ of the continuous variable (Fig. 6.20, right).

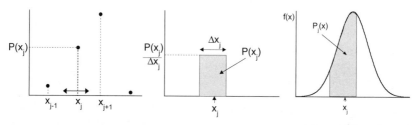

Fig. 6.20. Illustrating the comparison between a distribution of a discrete random variable (left) and a distribution of a continuous random variable (right).

(b) The previous procedure is simplified when the values of the discrete variable are equally spaced by Δx. The values $\mathcal{P}(x_j)/\Delta x$ are the heights of the histogram columns, and can be directly compared with the probability density $f(x)$ of the continuous variable. For the binomial and Poisson distributions, $\Delta x = 1$, and the comparison with a probability density is particularly simple (Figs. 6.21 and 6.22).

(c) The interval of possible values of the continuous variable is divided in subintervals, each one centered on a value x_j of the discrete variable, and of width Δx_j. Let $\mathcal{P}_j(x)$ be the value of the integral of the probability density $f(x)$ calculated within the jth subinterval. The values $\mathcal{P}_j(x)$ can be directly compared with the values $\mathcal{P}(x_j)$ of the discrete variable.

The Normal Distribution as a Limit of Other Distributions

As a consequence of the central limit theorem, the *binomial distribution* (6.2) tends to a distribution of normal shape when the number of trials $n \to \infty$, the mean np remaining finite. A graphical illustration is given in Fig. 6.21. In Sect. 6.1, it was also shown that the skewness and kurtosis coefficients of the binomial distribution tend to zero when $n \to \infty$. An interesting application has been given in Example 6.17.

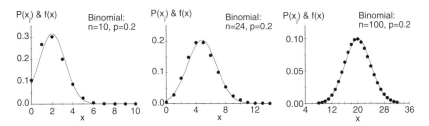

Fig. 6.21. Comparison among three binomial distributions $\mathcal{P}(x_j)$ and the corresponding normal distributions $f(x)$ with the same mean and variance. The skewness of the binomial distribution with $p = 0.2$ progressively decreases when n increases. The binomial variable has equally spaced values, $\Delta x = 1$, and it is possible to represent both the distribution $\mathcal{P}(x_j)$ and the probability density $f(x)$ on the same vertical axis.

As a consequence again of the central limit theorem, the *Poisson distribution* (6.45) tends to a distribution of normal shape when the parameter $a \to \infty$ (remember that a is both mean and variance of the Poisson distribution). A graphical illustration is given in Fig. 6.22. In Sect. 6.4, it was also shown that the skewness and kurtosis coefficients of the Poisson distribution tend to zero when $a \to \infty$.

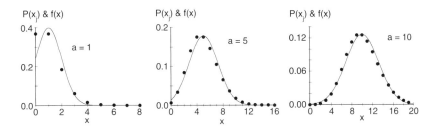

Fig. 6.22. Comparison among three Poisson distributions $\mathcal{P}(x_j)$ and the corresponding normal distributions $f(x)$ with the same mean and variance. The skewness of the Poisson distribution progressively decreases when a increases. The Poisson variable has equally spaced values, $\Delta x = 1$, and it is possible to represent both the distribution $\mathcal{P}(x_j)$ and the probability density $f(x)$ on the same vertical axis.

The probability calculations for the binomial and Poisson distributions can become rather lengthy when the parameters n and a, respectively, become large. In these cases, it can be convenient to substitute the binomial or Poisson distributions with the normal distribution of the same mean and variance, taking care to check that the substitution guarantees the required degree of accuracy.

Example 6.18. A coin is tossed $n = 10$ times, and one seeks the probability of obtaining a number of heads included between $k = 3$ and $k = 6$. The exact solution is given by the binomial distribution with $p = q = 0.5$:

$$\mathcal{P}(3 \leq k \leq 6) = \sum_{k=3}^{6} \binom{10}{k} 0.5^{10} = 0.7734 \, .$$

An approximate solution can be obtained by using the normal distribution $f(x)$ with the same mean $m = np = 5$ and variance $\sigma^2 = npq = 2.5$. Attention has to be paid to the transition from the discrete to the continuous variable. The area of a histogram column has to be substituted to each discrete value $\mathcal{P}(k_j)$. The base of the first histogram column, centered at $k = 3$, begins at $k = 2.5$; the base of the last column, centered at $k = 6$, ends at $k = 6.5$. Using the normal distribution, the sought probability is the integral of the density $f(x)$ calculated from $x_\alpha = 2.5$ to $x_\beta = 6.5$, that corresponds to the integral of the standard density from $z_\alpha = -1.58$ to $z_\beta = 0.95$. From the tables of Appendix C.3, one finds:

$$\mathcal{P}(3 \leq k \leq 6) = \int_{2.5}^{6.5} f(x) \, dx = \int_{-1.58}^{0.95} \phi(z) \, dz = 0.7718 \, .$$

Other examples of distributions that asymptotically tend to the normal distribution are encountered in the next chapters.

6.7 The Cauchy–Lorentz Distribution

The Cauchy distribution (after the name of the French mathematician and engineer A. L. Cauchy, 1798–1857) is interesting for its particular mathematical properties. It finds many applications in physics, where it is more frequently known as the Lorentz distribution (after the name of the Dutch physicist H. Lorentz, 1853–1928) or also the Breit–Wigner distribution.

The Cauchy–Lorentz (or Breit–Wigner) distribution is a distribution of a continuous random variable, whose density is expressed by the function

$$f(x) = \frac{1}{\pi} \frac{\gamma}{(x - \mu)^2 + \gamma^2} \, . \tag{6.80}$$

The distribution depends on two parameters, μ and γ, and one can show that it is normalized to one (Appendix D.7). The distribution (Fig. 6.23,

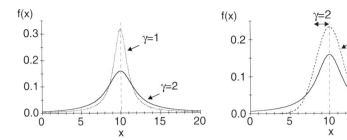

Fig. 6.23. Left: two Cauchy–Lorentz distributions with equal mode $\mu = 10$ and different half-width, $\gamma = 1$ and $\gamma = 2$. Right: comparison between a Cauchy–Lorentz distribution and a normal distribution with equal mode $\mu = 10$ and half-width at half maximum $\gamma = 2$ (corresponding, for the normal distribution, to $\sigma = 1.709$).

left) is symmetrical with respect to $x = \mu$, and asymptotically goes to zero for $x \to \pm\infty$. The value $x = \mu$ corresponds to the mode of the distribution.

The Cauchy–Lorentz distribution has certain pathological properties. In fact, one cannot define the variance, because the corresponding integral does not converge; for subtler reasons, one cannot define the mean as well (see Appendix D.7 for more details). The position of the distribution is thus defined by the mode μ (which is sometimes improperly called the mean). To measure the dispersion of the distribution one refers to the full-width at half maximum Γ. One can easily evaluate the density (6.80) in correspondence of $x = \mu$ and $x = \mu + \gamma$:

$$f(\mu) = \frac{1}{\pi\gamma}, \qquad f(\mu+\gamma) = \frac{1}{2\pi\gamma} = \frac{1}{2} f(\mu). \qquad (6.81)$$

The parameter γ represents the half-width at half maximum, so that $\Gamma = 2\gamma$.

The Cauchy–Lorentz distribution is intrinsically different from the normal distribution, as one can verify by comparing the two distributions with the same values of mode and half-width at half maximum (Fig. 6.23, right): the Cauchy–Lorentz distribution is lower in correspondence to the mode, but is higher at the tails (this is basically the reason why the variance integral does not converge).

Some important applications of the Cauchy–Lorentz distribution in physics are illustrated by the following examples.

Example 6.19. A radioactive sample is in position P (Fig. 6.24), at a distance s from a row of detectors placed along the x-axis. The decay products are emitted by the radioactive sample in an isotropic way, so that the probability density as a function of the angle θ is constant: $\phi(\theta) = C$. To determine the probability density $f(x)$ as a function of the position x along the detector row, one first notices that the variables θ and x are connected by

$$x = s\,\mathrm{tg}\,\theta, \qquad \mathrm{d}x = \frac{\mathrm{d}x}{\mathrm{d}\theta}\,\mathrm{d}\theta = \frac{s}{\cos^2\theta}\,\mathrm{d}\theta = \frac{x^2+s^2}{s}\,\mathrm{d}\theta\,.$$

The probability elements must be equal, $\phi(\theta)\,d\theta = f(x)\,dx$, therefore one gets

$$f(x) = \phi(\theta)\frac{d\theta}{dx} = C\frac{s}{x^2 + s^2}\,. \tag{6.82}$$

The probability density $f(x)$ has a Cauchy–Lorentz shape (6.80), with $\mu = 0$, $\gamma = s$, and $C = 1/\pi$.

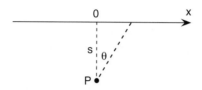

Fig. 6.24. Illustrating Example 6.19.

Example 6.20. Many physical systems can be described as forced and damped harmonic oscillators; the power P absorbed by the oscillator depends on the excitation frequency according to a law that, for small enough damping constant γ, has a Lorentzian shape

$$P(\omega) \propto \frac{\gamma}{(\omega - \omega_0)^2 + \gamma^2}\,, \tag{6.83}$$

where ω_0 is the proper frequency. For $\omega = \omega_0$, one observes the resonance phenomenon. In this example, Equation (6.83) describes a deterministic behavior, and is not a probability distribution. Other phenomena of physics are governed by the resonance equation (6.83) where, however, $P(\omega)$ has the meaning of a probability density. For example, the quantum emission of electromagnetic radiation by excited atoms is a random phenomenon, where the frequency ω of the emitted radiation is randomly distributed around a central frequency ω_0 according to (6.83).

6.8 Multivariate Distributions

In the previous sections, attention has been focused on random phenomena that can be described by one random variable. There are, however, many cases where the use of two or more random variables is required. The distributions of two or more random variables are called *multivariate distributions*.

Example 6.21. The measurement of a physical quantity is repeated N times, with sufficiently low resolution ΔX to allow the dispersion due to random fluctuations. The result x_i of a single measurement is a random variable, whose limiting distribution $f(x)$ is generally normal. In some cases, it is useful to collectively consider the set of all N results, and seek the probability density of a given N-fold of values $x_1, x_2, x_3, \ldots, x_N$. This problem leads to a multivariate distribution with respect to N random variables.

The treatment of multivariate distributions is by far more complex than the treatment of univariate distributions. In this section, attention is limited to a few basic concepts, necessary for the applications of subsequent chapters. Only distributions of two random variables (bivariate distributions) are considered, the extension to multivariate distributions being straightforward.

Multivariate Distributions for Discrete Random Variables

Let us consider two discrete random variables, H and K, whose possible values are $h_1, h_2, \ldots, h_i, \ldots$, and $k_1, k_2, \ldots, k_j, \ldots$, respectively.

The bivariate distribution is represented by the values of probability for all possible pairs (h_i, k_j) of values of the variables H and K.

Table 6.2. Graphical representation of a bivariate distribution for discrete random variables.

	h_1	h_2	$\ldots$	h_i	$\ldots$	$\sum_i$
k_1	$\mathcal{P}(h_1, k_1)$	$\mathcal{P}(h_2, k_1)$	$\ldots$	$\mathcal{P}(h_i, k_1)$	$\ldots$	$\mathcal{P}_k(k_1)$
k_2	$\mathcal{P}(h_1, k_2)$	$\mathcal{P}(h_2, k_2)$	$\ldots$	$\mathcal{P}(h_i, k_2)$	$\ldots$	$\mathcal{P}_k(k_2)$
$\ldots$	$\ldots$	$\ldots$	$\ldots$	$\ldots$	$\ldots$	$\ldots$
k_j	$\mathcal{P}(h_1, k_j)$	$\mathcal{P}(h_2, k_j)$	$\ldots$	$\mathcal{P}(h_i, k_j)$	$\ldots$	$\mathcal{P}_k(k_j)$
$\ldots$	$\ldots$	$\ldots$	$\ldots$	$\ldots$	$\ldots$	
$\sum_j$	$\mathcal{P}_h(h_1)$	$\mathcal{P}_h(h_2)$	$\ldots$	$\mathcal{P}_h(h_i)$	$\ldots$	1

A bivariate distribution for discrete random variables can be represented as in Table 6.2. The first row and the first column contain the values h_i and k_i of the two variables, respectively. The central part contains the probabilities $\mathcal{P}(h_i, k_j)$ for all pairs of values (h_i, k_j). The normalization condition over all possible pairs of values must hold:

$$\sum_i \sum_j \mathcal{P}(h_i, k_j) = 1 . \tag{6.84}$$

The probability that the variable H has a given value h_i, independently of the values of the variable K, is given by the sum over all values of K:

$$\mathcal{P}_h(h_i) = \sum_j \mathcal{P}(h_i, k_j) . \tag{6.85}$$

Similarly, the probability that the variable K has a given value k_j, independently of the values of the variable H, is given by the sum over all values of H:

$$\mathcal{P}_k(k_j) = \sum_i \mathcal{P}(h_i, k_j) \,. \tag{6.86}$$

The values $\mathcal{P}_h(h_i)$ (6.85) and $\mathcal{P}_k(k_j)$ (6.86) are generally listed in the last line (inferior margin) and last column (right margin) of the table, respectively, as in Table 6.2. For that reason, the probabilities $\mathcal{P}_h(h_i)$ and $\mathcal{P}_k(k_j)$ are called *marginal probabilities*.

If all values $\mathcal{P}(h_i, k_j)$ are known, it is possible to calculate all the marginal probabilities $\mathcal{P}_h(h_i)$ and $\mathcal{P}_k(k_j)$ by (6.85) and (6.86), respectively. Conversely, if the marginal probabilities $\mathcal{P}_h(h_i)$ and $\mathcal{P}_k(k_j)$ are known, it is generally impossible to recover the distribution $\mathcal{P}(h_i, k_j)$, unless the random variables H and K are independent (see below).

Multivariate Distributions for Continuous Random Variables

For two continuous random variables X and Y, the bivariate distribution is generally expressed by a *double probability density*, say a function $f(x, y)$ of two variables.

The probability that the random variable X has a value included in the interval a–b and the random variable Y has a value included in the interval c–d is given by the double integral

$$\mathcal{P}(a < X < b,\ c < Y < d) = \int_{x=a}^{b} \int_{y=c}^{d} f(x, y)\, \mathrm{d}x\, \mathrm{d}y \,. \tag{6.87}$$

The normalization condition must hold:

$$\iint_{-\infty}^{+\infty} f(x, y)\, \mathrm{d}x\, \mathrm{d}y = 1 \,. \tag{6.88}$$

Also for continuous variables, as for discrete variables, one can define the marginal probabilities. The probability density for a given value x of the random variable X, independent of the value of the variable Y, is

$$f_x(x) = \int_{-\infty}^{+\infty} f(x, y)\, \mathrm{d}y \,. \tag{6.89}$$

Similarly, the probability density for a given value y of the random variable Y, independent of the value of the variable X, is

$$f_y(y) = \int_{-\infty}^{+\infty} f(x, y)\, \mathrm{d}x \,. \tag{6.90}$$

If the double density $f(x, y)$ is known, one can calculate the marginal densities $f_x(x)$ and $f_y(y)$ through (6.89) and (6.90), respectively. Conversely, if the marginal densities $f_x(x)$ and $f_y(y)$ are known, it is generally impossible to recover the double density $f(x, y)$, unless the variables X and Y are independent.

Independent Random Variables

Two random variables are said to be independent if the distribution law of one of them does not depend on the value of the other, and viceversa. If two random variables are independent, it is possible to factorize their bivariate distribution as a product of the marginal distributions.

For discrete random variables

$$\mathcal{P}(h_i, k_j) = \mathcal{P}_h(h_i)\mathcal{P}_k(k_j) \quad \text{(independent r.v.)}. \tag{6.91}$$

For continuous random variables

$$f(x,y) = f_x(x) f_y(y) \quad \text{(independent r.v.)}. \tag{6.92}$$

Numerical Characteristics, Covariance

As for univariate distributions, also for multivariate distributions one can define numerical characteristics, such as initial and central moments. Only the lowest-order moments of bivariate distributions are defined here, and only continuous random variables are considered, the extension to discrete random variables being trivial.

The *means* of the two variables X and Y are defined as

$$m_x = \langle x \rangle = \iint_{-\infty}^{+\infty} x\, f(x,y)\, \mathrm{d}x\, \mathrm{d}y\,, \tag{6.93}$$

$$m_y = \langle y \rangle = \iint_{-\infty}^{+\infty} y\, f(x,y)\, \mathrm{d}x\, \mathrm{d}y\,. \tag{6.94}$$

Calculating the mean of each variable requires knowledge of the bivariate density $f(x,y)$ and an integral over the two-dimensional space. If, however, X and Y are independent, one can factorize the density $f(x,y)$ according to (6.92), and, taking advantage of the normalization condition, express the means in terms of the one-dimensional marginal densities:

$$m_x = \int_{-\infty}^{+\infty} x\, f_x(x)\, \mathrm{d}x \quad \text{(independent r.v.)}, \tag{6.95}$$

$$m_y = \int_{-\infty}^{+\infty} y\, f_y(y)\, \mathrm{d}y \quad \text{(independent r.v.)}. \tag{6.96}$$

The *variances* of the two variables X and Y are defined as

$$D_x = \sigma_x^2 = \langle (x - m_x)^2 \rangle = \iint_{-\infty}^{+\infty} (x - m_x)^2\, f(x,y)\, \mathrm{d}x\, \mathrm{d}y\,, \tag{6.97}$$

$$D_y = \sigma_y^2 = \langle (y - m_y)^2 \rangle = \iint_{-\infty}^{+\infty} (y - m_y)^2\, f(x,y)\, \mathrm{d}x\, \mathrm{d}y\,. \tag{6.98}$$

Again, if X and Y are independent, one can express the variances in terms of the one-dimensional marginal densities:

$$D_x = \sigma_x^2 = \int_{-\infty}^{+\infty} (x - m_x)^2 \, f_x(x) \, dx \quad \text{(independent r.v.)}, \quad (6.99)$$

$$D_y = \sigma_y^2 = \int_{-\infty}^{+\infty} (y - m_y)^2 \, f_y(y) \, dy \quad \text{(independent r.v.)}. \quad (6.100)$$

When dealing with two random variables, mixed parameters can also be defined, involving the averages over both random variables. The lowest order of such mixed parameters is the *covariance*, say the average value of the product of the deviations of both random variables with respect to their means:

$$\sigma_{xy} = \langle (x - m_x)(y - m_y) \rangle$$
$$= \iint_{-\infty}^{+\infty} (x - m_x)(y - m_y) \, f(x,y) \, dx \, dy. \quad (6.101)$$

If X and Y are independent, making use of (6.92) one can factorize the integral (6.101) into a product of two integrals over the two marginal densities:

$$\sigma_{xy} = \int_{-\infty}^{+\infty} (x - m_x) \, f_x(x) \, dx \int_{-\infty}^{+\infty} (y - m_y) \, f_y(y)$$
$$= \langle (x - m_x) \rangle \langle (y - m_y) \rangle \quad \text{(independent r.v.)}. \quad (6.102)$$

According to (6.33), the central moment of order one of a univariate random variable is zero,

$$\langle (x - m_x) \rangle = 0, \quad \langle (y - m_y) \rangle = 0, \quad (6.103)$$

so that the covariance of two independent variables is zero, $\sigma_{xy} = 0$.

Important applications of the covariance are found in Chaps. 8 and 10.

Problems

6.1. *Binomial distribution.* Calculate and plot the binomial distributions for $n = 3, p = 0.3$, and for $n = 9, p = 0.3$.

6.2. *Binomial distribution.* An important application of the binomial distribution is the *random walk problem*. Let us consider a body moving along an x-axis at discrete steps of length d, starting from $x = 0$. The direction of each step (right or left) is random and independent of the direction of previous steps. Let p be the probability that one step is towards the right, and $q = 1 - p$ the probability that it is towards the left.

(a) Find the expression for the probability that k steps, out of n, are towards the right.

(b) Demonstrate that the net displacement towards the right after n steps is
$$L = kd - (n-k)d = (2k-n)d.$$

6.3. Let us consider a sequence of repeated independent trials of a random phenomenon, and let p be the probability of positive outcome, and $q = 1-p$ the probability of negative outcome. The number k of trials that are necessary for obtaining the first positive outcome is a discrete random variable that can assume a countably infinite number of values.

(a) Verify that the distribution law is $\mathcal{P}(k) = q^{k-1}p$, and plot the distribution for the values $p = 1/2$ and $p = 1/6$.
(b) Verify that the distribution law satisfies the normalization condition $\sum_{k=1}^{\infty} pq^{k-1} = 1$.

Hints: Substitute $p = 1 - q$ and $k - 1 = s$, and remember that $\sum_{s=0}^{\infty} q^s = 1/(1-q)$.

6.4. *Binomial distribution.* A coin is tossed n times, the probability of "head" is $p = 0.5$ for each toss. The probability of k heads out of n tosses is given by the binomial distribution. Calculate the numerical parameters of the distribution (mean m, standard deviation σ, relative width σ/m, skewness coefficient β, and kurtosis coefficient γ_2) for $n = 10$, $n = 50$, and $n = 100$, and verify their trends as a function of n.

6.5. *Binomial distribution.* Let us again consider the one-dimensional random walk of Problem 6.2. Each step has length d and probability p of being made towards the right, and $q = 1-p$ towards the left. The probability that k steps, out of n, are made towards the right is given by the binomial distribution, and the net displacement towards the right is a random variable $L = kd - (n-k)d = (2k-n)d$.

(a) Verify that the mean and variance of L are $m_L = nd(p-q)$ and $D_L = 4d^2npq$, respectively.
(b) Consider the case $p = q = 0.5$, and study the behavior of the mean m_L and the standard deviation σ_L as a function of the number n of steps.

6.6. *Poisson distribution.* Calculate and plot the Poisson distributions for $a = 2$ and for $a = 10$.

6.7. *Poisson distribution.* In a book of 500 pages, $n = 300$ printing errors are randomly distributed. The probability that an error, chosen by chance, is within a given page is thus $p = 1/500$. Calculate the probability that a given page contains $k = 2$ errors using both the binomial distribution and its approximation given by the Poisson distribution.

6.8. *Poisson distribution.* A harmful bacterium is present in air with the average density $\lambda = 100$ bacteria per cube meter. Calculate the probability that a sample volume $v = 2\,\mathrm{dm}^3$ contains at least one bacterium.

6.9. *Normal distribution.* A continuous random variable x is distributed according to a normal law, with mean $m = 3.5$ and standard deviation $\sigma = 1.7$. Making use of the tables of Appendix C.3, calculate: (a) the probability that $1 < x < 3.5$, and (b) the probability that $2 < x < 5$.

6.10. *Central limit theorem.* Let us consider a pair of vectors of unit magnitude $|\boldsymbol{u}| = |\boldsymbol{v}| = 1$, aligned along the same z direction, and let $Y = |\boldsymbol{u}+\boldsymbol{v}|$ be the magnitude of their sum. The two vectors have equal probability $\mathcal{P} = 0.5$ of being parallel ($\uparrow\uparrow$, $Y = 2$) or antiparallel ($\uparrow\downarrow$, $Y = 0$.) (A physical example is given by two paired electrical dipoles).

Let us now consider a set of n independent pairs, and the sum $S = \sum Y_i$, where $Y_i = 0$ or $Y_i = 2$. S is a random variable that can assume nonnegative integer even values. Calculate the distribution of the random variable S, the mean m, and the variance D for selected values of n, and graphically check that the shape of the distribution tends to the normal shape when n increases.

7 Statistical Tools

In Chap. 4, when dealing with measurements affected by random fluctuations (Sect. 4.3), the abstract concept of limiting distribution was introduced. Some procedures for estimating the parameters of the limiting distribution from the results of a finite number of measurements were also presented, without, however, a rigorous justification.

After the introduction of the basic concepts of probability theory and of the distributions of random variables (Chaps. 5 and 6), it is now possible to give more sound foundations to those procedures, based on statistical methods.

7.1 Parent and Sample Populations

The concepts of *population* and *sample* are basic to statistics. In this section, their meaning is clarified, with particular attention to the applications to the data analysis procedures.

Parent Population

Populations of a *finite* number of individuals are familiar to social sciences and economics. If one focuses attention on a given property of a population (such as the age of a group of persons or the length of a set of screws) one can build up a *parent population* relative to that property. If the property can be described by a random variable, the parent population can be represented by a distribution of the random variable.

Example 7.1. The electronic circuits produced by a factory in a month represent a finite population, each circuit being an individual. The working time of each circuit is a random variable. The parent population relative to the working time can be determined by monitoring all circuits. Once the parent population is known, one can determine the probability density that a circuit, chosen by chance, breaks down after a given time interval.

For the applications of interest in this book, one has to deal with abstract parent populations, containing an *infinite* number of individuals. Let us clarify this concept with some examples.

Example 7.2. In a random phenomenon, a given outcome (such as obtaining "5" when a die is tossed) has the probability p, and $q = 1-p$ is the probability of the alternative outcomes. One describes the phenomenon by a discrete random variable with two values: $k_1 = 0$ for the negative outcomes, $k_2 = 1$ for the positive outcome. Each repetition of the experiment is an individual, and the parent population is made up of all the infinite possible repetitions of the experiment. The distribution of the parent population is the binomial distribution (6.2) with $n = 1$: $\mathcal{P}(k_1) = q$, $\mathcal{P}(k_2) = p$. Mean and variance are $m = p$ and $D = pq$, respectively.

Example 7.3. A random phenomenon consists of n repetitions of the same trial, with probability p of positive outcome. One can describe the phenomenon by a discrete random variable K, which counts the trials with positive outcome. Each sequence of n trials is now an individual, and the parent population is made up of all the infinite possible repetitions of n trials. The distribution of the parent population is the binomial distribution (6.2).

Example 7.4. The result of the measurement of a physical quantity affected by random fluctuations is a random variable. The parent population consists of all the infinite possible measurements. The distribution of the parent population can often be approximated by a normal distribution.

Example 7.5. The number of cosmic rays detected by a Geiger counter in one minute is a discrete random variable. Each measurement lasting one minute is an individual, and the parent population consists of all the infinite possible measurements lasting one minute. The distribution of the parent population is the Poisson distribution.

Sample Population

The experimental determination of the properties of a parent population can be exceedingly time consuming and expensive for large finite populations (Example 7.1), and it is impossible for infinite populations (Examples 7.2–7.5). In practice, one selects and examines a limited number N of individuals of the parent population, say a statistical sample. The sampling, say the choice of the sample, must guarantee that the sample is a good representative of the entire population. The problem of representativeness of the sample can be very complex in social sciences and economics; it is generally simpler in physics, where a necessary condition for representativeness of the sample is the randomness of its choice, and the larger is the size of the sample, the more representative it is.

The distribution of values of the random variable within the sample is called the *sample distribution*. The sample distribution has a random character, because it depends on the sample.

The characteristics of the parent population are estimated from the random characteristics of the sample population through the procedure of *statistical inference*. Let us now reconsider the previous examples.

Example 7.6. (See Example 7.1) To estimate the average lifetime of the electronic circuits produced in a month by a factory, one randomly chooses a small number N of them: they represent the sample that will be tested for lifetime.

Example 7.7. (See Example 7.2) For a single random phenomenon, the sample is represented by N identical repetitions. The outcome is negative in n_1^* repetitions; it is positive in n_2^* repetitions $(n_1^* + n_2^* = N)$. The ratios $p^* = n_2^*/N$ and $q^* = n_1^*/N$ are the frequencies of realization and nonrealization, respectively. The sample distribution consists of the two values $\mathcal{P}^*(k_2) = p^*$ and $\mathcal{P}^*(k_1) = q^*$. The frequencies p^* and q^* have a random character. When the sample size increases ($N \to \infty$), one expects that the sample frequencies tend to the probabilities, $p^* \to p$, $q^* \to q$ (Fig. 7.1).

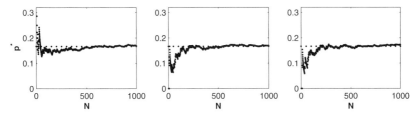

Fig. 7.1. Sampling of a die toss. The parent population is represented by the infinite possible tosses, and the parent distribution consists of two values of probability, $p = 1/6$ (dashed line) for a given outcome, and $q = 5/6$ for the alternative outcomes. The three graphs refer to three different sampling sequences, and show the sample frequency p^* as a function of the sample size N. When N increases, the statistical frequency p^* stabilizes around the probability p.

Example 7.8. Let us consider again the binomial distribution (Example 7.3). A sample is represented by N independent repetitions of the sequence of n single trials. At each repetition, the random variable K has a given value k. For N repetitions, a given value k is obtained $n^*(k)$ times. The sample distribution is represented by the values of the sample frequencies, $\mathcal{P}^*(k) = n^*(k)/N$. When N increases ($N \to \infty$), the sample distribution approaches the binomial parent population, $\mathcal{P}^*(k) \to \mathcal{P}(k)$ (Fig. 7.2).

Example 7.9. Let us now reconsider the measurement of a physical quantity (Example 7.4). The N experimental values are a sample of the parent population represented by the outcomes of infinite possible measurements. The sample distribution is generally represented by a histogram with $\mathcal{N}$ columns,

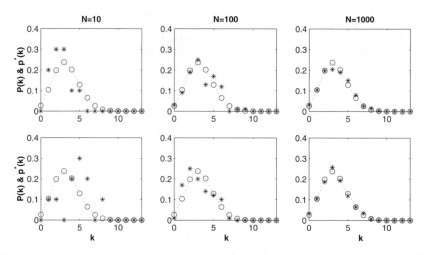

Fig. 7.2. Sampling of a binomial distribution relative to $n = 20$ tosses of a die ($p = 1/6$). The parent distribution is represented by open circles. The sample populations (asterisks) refer to samples of different sizes: $N = 10$ (left); $N = 100$ (center); $N = 1000$ (right). The random difference between different samples (top and bottom graphs) reduces when N increases.

whose areas are proportional to the sample frequencies $p_j^* = n_j^*/N$. When the sample size increases ($N \to \infty$), the area of each column approaches the area of the limiting distribution $f(x)$ within the corresponding interval Δx_j: $p_j^* \to p_j = \int f(x)\,dx$ (Fig. 7.3).

Example 7.10. Finally, let us reconsider the Poisson distribution (Example 7.5). A sample is represented by N independent repetitions of the counting for one minute. At each repetition, the random variable K assumes a well-defined value. For N repetitions, a given value k is obtained $n^*(k)$ times. The sample distribution is given by the values of the sample frequencies $\mathcal{P}^*(k) = n^*(k)/N$. When the sample size increases ($N \to \infty$) the sample distribution is expected to approach the Poisson parent distribution: $\mathcal{P}^*(k) \to \mathcal{P}(k)$.

In the following, we consider only samples of infinite parent populations, such as those in Examples 7.7 through 7.10. When the sample size increases, the sample distributions tend to stabilize and to become more and more similar to the parent distributions. The convergence of the sample distributions to the parent distributions has a random character, and is codified by a group of theorems of probability theory, that are globally referred to as the *law of large numbers*. In particular, the law of large numbers describes the convergence of the statistical frequency of an event to its probability, on which the statistical evaluation of probability is based (Sect. 5.3).

When finite parent populations are sampled, as in Example 7.6, it is necessary to distinguish between samplings with and without reposition. Some

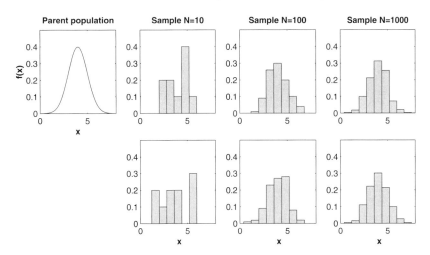

Fig. 7.3. Parent normal distribution with $m = 4$ and $\sigma = 1$ (top left) and six different sample distributions corresponding to samples of different size, $N = 10$ (center-left), $N = 100$ (center-right), $N = 1000$ (right).

properties of the samples of infinite populations can be extended to the samples of finite populations only if the sampling is done with reposition.

7.2 Sample Means and Sample Variances

Position and dispersion of a parent distribution can be measured by two parameters, the parent mean (or expected value) m and the parent variance D, that are defined (Sect. 6.3) as

$$m = \sum_j x_j p_j \,, \quad D = \sum_j (x_j - m)^2 p_j \tag{7.1}$$

for discrete populations, and

$$m = \int_{-\infty}^{+\infty} x\, f(x)\, dx \,, \quad D = \int_{-\infty}^{+\infty} (x - m)^2 f(x)\, dx \tag{7.2}$$

for continuous populations.

A statistical sample of size N consists of N values $x_1, x_2, \ldots, x_N$ of the random variable associated with the population. The sample mean m^* and the sample variance D^* are defined as arithmetic averages:

$$m^* = \frac{1}{N} \sum_{i=1}^{N} x_i \,, \quad D^* = \frac{1}{N} \sum_{i=1}^{N} (x_i - m^*)^2 \,. \tag{7.3}$$

Alternatively, the sample mean and the sample variance can be expressed as a function of the statistical frequencies p_j^*, where the index j labels the different values x_j for discrete populations (Fig. 7.2) or the columns of a histogram for continuous populations (Fig. 7.3):

$$m^* = \sum_{j=1}^{\mathcal{N}} x_j p_j^*, \quad D^* = \sum_{j=1}^{\mathcal{N}} (x_j - m^*)^2 p_j^*. \quad (7.4)$$

The formal similarity between (7.4) and (7.1) enlightens the relation between statistical frequencies p^* and probabilities p.

Limiting Distributions of Sample Means and Variances

The sample mean m^* and the sample variance D^* are continuous random variables, whose values depend on the particular sample. By repeating the sampling many times, one can build up the histograms of sample means and sample variances.

Example 7.11. Let us consider the toss of a die (Example 7.2). The probability of the outcome "5" is $p = 1/6$. The outcome of the toss can be described by two values of a random variable K: $k_1 = 0$ for outcomes different from "5", and $k_2 = 1$ for the outcome "5". The parent distribution of the random variable K has mean $m = p$ and variance $D = pq$. Figure 7.4 shows the histograms of the sample means m^* (left) and sample variances D^* (right) obtained from samples of $N = 50$ and $N = 500$ tosses. Each histogram is based on 100 repetitions of the sampling. It is evident that the width of the histograms is smaller for $N = 500$ than for $N = 50$.

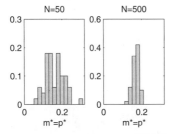
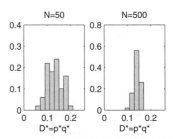

Fig. 7.4. Sampling the population relative to the toss of a die (Example 7.11). Two samples of $N = 50$ and $N = 500$ tosses are considered; for each sample, from the frequency p^* of the outcome "5", the sample mean $m^* = p^*$ and sample variance $D^* = p^*q^*$ are calculated. By repeating both sampling procedures 100 times, the histograms of sample means (left) and sample variances (right) have been obtained.

When the number of samples is increased, the histograms of sample means and sample variances tend to asymptotic shapes, the limiting distributions of sample means and sample variances.

7.2 Sample Means and Sample Variances

A limiting distribution of sample means and its main properties have been considered in Sect. 4.3, when treating the effect of random fluctuations on measurement results.

The properties of the limiting distributions of sample means and sample variances are now studied from a general point of view. We show that the numerical parameters of the distributions of sample means and sample variances are connected to the parameters of the parent distribution of single values by simple relations.

In laboratory practice, one generally performs only one sampling of a population, so that a single value of sample mean m^* and a single value of sample variance D^* are obtained. The properties of the limiting distributions studied in this section are used in Sect. 7.3 to estimate the parameters of a parent population from the results of a single sampling. The properties of the distributions of means are explored in Experiment E.2 of Appendix E.

Useful Relations

Let X and Y be two random variables, and a a real constant. The mean and variance are indicated here by $\mathbf{m}[\ldots]$ and $\mathbf{D}[\ldots]$, respectively. One can demonstrate (Appendix D.8) that:

$$\mathbf{m}[aX] = a\,\mathbf{m}[X], \tag{7.5}$$
$$\mathbf{D}[aX] = a^2\,\mathbf{D}[X], \tag{7.6}$$
$$\mathbf{m}[X+Y] = \mathbf{m}[X] + \mathbf{m}[Y], \tag{7.7}$$
$$\mathbf{D}[X+Y] = \mathbf{D}[X] + \mathbf{D}[Y]. \tag{7.8}$$

The first three relations (7.5) through (7.7) are always true, whereas (7.8) is only true if X and Y are independent variables. The equations (7.7) and (7.8) can be immediately generalized to the addition of more than two random variables.

Distribution of Sample Means

The *mean value* of the limiting distribution of sample means, $\mathbf{m}[m^*]$, is equal to the parent mean m. In fact, by using (7.5) and (7.7),

$$\begin{aligned}
\mathbf{m}[m^*] &= \mathbf{m}\left[\frac{X_1 + X_2 + \cdots + X_N}{N}\right] \\
&= \frac{1}{N}\{\mathbf{m}[X_1] + \mathbf{m}[X_2] + \cdots + \mathbf{m}[X_N]\} \\
&= \frac{1}{N}\{N\,m\} = m\,. \tag{7.9}
\end{aligned}$$

The *variance* of the limiting distribution of sample means, $\mathbf{D}[m^*]$, is N times smaller than the parent variance D. In fact, by using (7.6) and (7.8),

$$\mathbf{D}[m^*] = \mathbf{D}\left[\frac{X_1 + X_2 + \cdots + X_N}{N}\right]$$

$$= \frac{1}{N^2}\{\mathbf{D}[X_1] + \mathbf{D}[X_2] + \cdots + \mathbf{D}[X_N]\}$$

$$= \frac{1}{N^2}\{N D\} = \frac{D}{N}. \tag{7.10}$$

The limiting distribution of sample means has the same mean m of the parent distribution, and its variance decreases as $1/N$ when the sample size increases (Fig. 7.4, left). As a consequence of the central limit theorem (Sect. 6.6), the limiting distribution of sample means tends to the normal shape when the sample size increases.

These properties of the distribution of sample means have already been utilized in Sect. 4.3, when treating the uncertainty due to random fluctuations.

Distribution of Sample Variances

To calculate the *mean* of the limiting distribution of sample variances $\mathbf{m}[D^*]$, let us first consider the deviation of the ith value x_i from the sample mean m^*, and express it as a function of the deviation from the parent mean m:

$$(x_i - m^*) = (x_i - m) - (m^* - m), \tag{7.11}$$

so that

$$\sum_{i=1}^{N}(x_i - m^*)^2 = \sum_{i=1}^{N}(x_i - m)^2 - 2(m^* - m)\sum_{i=1}^{N}(x_i - m) + N(m^* - m)^2$$

$$= \sum_{i=1}^{N}(x_i - m)^2 - 2(m^* - m)N(m^* - m) + N(m^* - m)^2$$

$$= \sum_{i=1}^{N}(x_i - m)^2 - N(m^* - m)^2. \tag{7.12}$$

The sample variance is thus

$$D^* = \frac{1}{N}\sum_{i=1}^{N}(x_i - m^*)^2 = \frac{1}{N}\sum_{i=1}^{N}(x_i - m)^2 - (m^* - m)^2, \tag{7.13}$$

and the mean of the distribution of sample variances is

$$\mathbf{m}[D^*] = \mathbf{m}\left[\frac{1}{N}\sum_{i=1}^{N}(x_i - m)^2\right] - \mathbf{m}\left[(m^* - m)^2\right]. \tag{7.14}$$

The first term of the right-handside member is the mean of the squared deviation of the x_i values from the parent mean m, say the parent variance

D. The second term of the right-handside member is the mean of the squared deviation of the sample mean m^* from the parent mean m; according to (7.9), $m = \mathbf{m}[m^*]$, so that the second term corresponds to the variance of the distribution of sample means $\mathbf{D}[m^*]$, which in turn, according to (7.10), is $\mathbf{D}[m^*] = D/N$. In conclusion,

$$\mathbf{m}[D^*] = D - \frac{1}{N}D = \frac{N-1}{N}D. \tag{7.15}$$

The mean of the limiting distribution of sample variances is smaller than the parent variance D. The difference decreases when the sample size N increases.

It is worth noting that, in order to calculate the sample variance D^*, one must previously calculate the sample mean m^*. The N terms of the sum in (7.12) are thus not independent. The factor $N-1$ in (7.15) is the number of independent terms in the expression of variance, and is called the number of *degrees of freedom*.

7.3 Estimation of Parameters

One of the fundamental problems of statistics, so-called statistical inference, consists of getting the maximum amount of information on a parent distribution from a finite experimental sample. This problem, phenomenologically introduced in Sect. 4.3, is considered here from a rather general point of view.

The functional form of a parent population (normal, binomial, Poisson, etc.) is in many cases hypothesized from theoretical considerations or from experience in data analysis. In Chap. 11, a methodology is introduced that allows an a posteriori evaluation of the soundness of an hypothesis of functional form of a distribution.

Once the form of the distribution has been established, its numerical parameters (mean, variance, skewness coefficient, etc.) can be estimated from the experimental data that represent a sample population. In the following, a generic parameter of the parent distribution and of the sample distribution is indicated as λ and λ^*, respectively.

Example 7.12. The limiting distribution of the values of a physical quantity affected by random fluctuations can often be approximated by a normal distribution (Sect. 4.3). The parent parameters λ_1 and λ_2 are the parent mean and the parent variance, m and $D = \sigma^2$, respectively. The corresponding sample parameters λ_1^* and λ_2^* are the mean and the variance of the experimental histogram, m^* and D^*, respectively.

Example 7.13. The histogram representing the measured values of a given physical quantity has a bimodal shape. One hypothesizes that the physical quantity can assume two distinct values, and that each value is dispersed according to a normal distribution. One assumes a parent population

$$f(x) = A_1 \frac{1}{\sigma_1\sqrt{2\pi}} \exp\left[\frac{(x-m_1)^2}{2\sigma_1^2}\right] + A_2 \frac{1}{\sigma_2\sqrt{2\pi}} \exp\left[\frac{(x-m_2)^2}{2\sigma_2^2}\right],$$

whose six parameters are: $\lambda_1 = A_1$, $\lambda_2 = m_1$, $\lambda_3 = \sigma_1$, $\lambda_4 = A_2$, $\lambda_5 = m_2$, and $\lambda_6 = \sigma_2$.

The sample parameters λ_s^* are random variables, because they depend on the random values $x_1, x_2, \ldots, x_N$ of the sample. It is impossible to exactly determine the parent parameters from a finite sample. Some procedures have, however, been developed that allow an estimate of the parent parameters starting from the sample parameters.

An *estimator* of a parameter λ is indicated here by $\tilde{\lambda}$ (a similar convention has already been used in Chap. 4). Also the value of the estimator, say the estimate, is indicated by $\tilde{\lambda}$.

Properties of Estimators

Different procedures, say different estimators, can be devised to estimate a parameter λ from the N values of a sample. For any procedure, an estimate $\tilde{\lambda}$ is always a random variable whose value randomly changes when the sampling is iterated: one can thus speak of a *distribution of the estimates* $\tilde{\lambda}$. In order to establish effective criteria for the comparison and choice of different procedures for estimating parameters, one defines some properties that can be attributed to estimators. An estimator $\tilde{\lambda}$ is

(a) *Consistent*, if $\lim_{N\to\infty} \tilde{\lambda} = \lambda$, say if the estimate $\tilde{\lambda}$ tends to the parent parameter λ when the sample size increases

(b) *Unbiased*, if $\mathbf{m}[\tilde{\lambda}] = \lambda$, say if the mean of the limiting distribution of $\tilde{\lambda}$ corresponds to the parent parameter λ

(c) *Effective*, if $\mathbf{D}[\tilde{\lambda}]$ is minimum, say if the chosen estimator gives a distribution of estimates with the smallest variance with respect to all the alternative estimators of the same parameter λ

Example 7.14. Let us consider the mean of a parent population: $\lambda_1 = m$. A possible estimator of m, based on a sample of size N, is the sample average

$$\tilde{m} = m^* = \sum_i x_i/N \,. \tag{7.16}$$

According to (7.9), the estimator is consistent and unbiased; one can demonstrate that it is also effective.

Example 7.15. Let us now consider the variance of a parent population: $\lambda_2 = D$. A possible estimator of D, based on a sample of size N, is the sample variance $\tilde{D} = D^*$. This estimator is consistent, but, according to (7.15), is not unbiased, because the mean of the distribution of sample variances does not correspond to the parent variance D. A consistent and unbiased estimator is instead

$$\tilde{D} = \frac{N}{N-1} D^* . \tag{7.17}$$

Criterion of Maximum Likelihood

A method that is frequently used to estimate parent parameters from a finite sample is based on the criterion of maximum likelihood. According to this criterion, the best estimators of the parent parameters are those that maximize the probability of obtaining exactly the sample on which the estimate is based.

To better grasp the idea, let us focus on a continuous random variable whose parent distribution $f(x)$ depends on a given number of parameters λ_s.

Let us first make the abstract hypothesis that the parent population $f(x)$ and its parameters λ_s are perfectly known. In this case, the probability density would be known for any value of x. Consider now a sample of size N (e.g., N measurements of a physical quantity). The probability density of obtaining a well defined N-fold of values $x_1, x_2, \ldots, x_N$ is the multivariate distribution $g(x_1, x_2, \ldots, x_N)$ (Sect. 6.8), which is sometimes called the *likelihood function*. If the results of sampling are independent, the distribution $g(x_1, x_2, \ldots, x_N)$ can be factorized into the product of N univariate marginal distributions, all characterized by the same density $f(x)$:

$$g(x_1, x_2, \ldots, x_N) = f(x_1) f(x_2) \cdots f(x_N) . \tag{7.18}$$

Let us now consider a real case, where N sample values $x_1, x_2, \ldots, x_N$ are known, and one seeks an estimate of the unknown parent parameters λ_s. The likelihood function (7.18) now depends on the n parameters λ_s, the values x_i being known: $g(x_1, x_2, \ldots, x_N; \lambda_1, \lambda_2, \ldots, \lambda_n)$. According to the maximum likelihood criterion, the best estimate of the parameters λ_s is given by the values $\tilde{\lambda}_s$ that maximize the probability density of exactly obtaining the N-fold of values $x_1, x_2, \ldots, x_N$:

$$\max\ [g(x_1, x_2, \ldots, x_N; \lambda_1, \lambda_2, \ldots, \lambda_n)] \Rightarrow \tilde{\lambda}_1, \tilde{\lambda}_2, \ldots, \tilde{\lambda}_n . \tag{7.19}$$

For example, let us suppose that the parent distribution is normal,

$$f(x) = \frac{1}{\sigma\sqrt{2\pi}} \exp\left[-\frac{(x-m)^2}{2\sigma^2}\right] , \tag{7.20}$$

and search for the best estimate of the parameters $\lambda_1 = m$ and $\lambda_2 = \sigma$ from a sample of N values $x_1, x_2, \ldots, x_N$. The probability density of obtaining the sample values $x_1, x_2, \ldots, x_N$ is given by the likelihood function

$$g(x_1, x_2, \ldots, x_N; m, \sigma) = \frac{1}{\sigma^N (\sqrt{2\pi})^N} \exp\left[-\sum_{i=1}^{N} \frac{(x_i - m)^2}{2\sigma^2}\right] , \tag{7.21}$$

where $x_1, x_2, \ldots, x_N$ are known, and m and σ are the unknowns. The best estimates $\tilde{m}$ and $\tilde{\sigma}$ of m and σ are the values that maximize the multivariate density $g(x_1, x_2, \ldots, x_N; m, \sigma)$. To evaluate them, let us impose that the first partial derivatives with respect to the unknowns are zero,

$$\frac{\partial g(x_1, \ldots, x_N; m, \sigma)}{\partial m} = 0, \qquad \frac{\partial g(x_1, \ldots, x_N; m, \sigma)}{\partial \sigma} = 0, \qquad (7.22)$$

and the second partial derivatives are negative. One can easily verify that:

$$\tilde{m} = \frac{1}{N} \sum_i x_i = m^*, \qquad \tilde{\sigma} = \sqrt{\frac{1}{N} \sum_i (x_i - m)^2}. \qquad (7.23)$$

Let us now compare (7.23) with (7.16) and (7.17). One can easily verify that the criterion of maximum likelihood gives an unbiased estimator of the mean $\tilde{m} = m^*$. The estimate $\tilde{\sigma}$ of (7.23) is based on the deviations with respect to the unknown parameter m; if m is substituted by its estimate m^*, one obtains an estimator of σ consistent but not unbiased. An unbiased estimator of $D = \sigma^2$ is given by (7.17).

Weighted Average

The weighted average was used in Sect. 4.4 to synthesize two or more consistent values of the same physical quantity obtained from different measurements. The weighted average procedure can now be derived from the criterion of maximum likelihood.

Let us consider the results of two sets of measurements of a physical quantity,

$$X_A \pm \delta X_A, \qquad X_B \pm \delta X_B, \qquad (7.24)$$

respectively, and suppose that:

(a) The results are consistent, according to the prescriptions of Sect. 4.4.
(b) The uncertainties are due to random fluctuations and expressed as standard deviations of the (normal) distributions of the sample means:

$$\delta X_A = \sigma_A[m_A^*], \qquad \delta X_B = \sigma_B[m_B^*]. \qquad (7.25)$$

The two results (7.24) should now be synthesized into a unique expression $X = X_0 \pm \delta X_0$. To derive the procedure of weighted average from the criterion of maximum likelihood, let us rely on the same approach as the previous subsection.

If the true value X_v of the physical quantity were known (abstract hypothesis), the probability densities of obtaining the central values X_A and X_B by the sampling procedures A and B, respectively, would be

7.3 Estimation of Parameters

$$f_A(X_A) = \frac{1}{\sigma_A\sqrt{2\pi}} \exp\left[-\frac{(X_A - X_v)^2}{2\sigma_A^2}\right], \quad (7.26)$$

$$f_B(X_B) = \frac{1}{\sigma_B\sqrt{2\pi}} \exp\left[-\frac{(X_B - X_v)^2}{2\sigma_B^2}\right]. \quad (7.27)$$

Equations (7.26) and (7.27) express the limiting distributions of the sample means (to simplify the notation, here σ stands for $\sigma[m^*]$): the distributions are both centered on X_v, but their widths depend on the different measurement procedures. We can consider (7.26) and (7.27) as two distinct parent distributions, relative to the measurements with the two procedures A and B, respectively (Fig. 7.5). The probability density of contemporarily obtaining the value X_A with the procedure A and the value X_B with the procedure B would be the product of $f_A(X_A)$ and $f_B(X_B)$, say the multivariate density:

$$g(X_A, X_B) = \frac{1}{\sigma_A\sigma_B\, 2\pi} \exp\left[-\frac{(X_A - X_v)^2}{2\sigma_A^2} - \frac{(X_B - X_v)^2}{2\sigma_B^2}\right]. \quad (7.28)$$

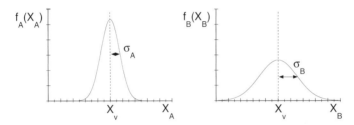

Fig. 7.5. Parent distributions relative to the measurements of the same quantity X by two different procedures, A and B, differently affected by random fluctuations.

Let us now consider the real situation, where the values X_A and X_B are known and X_v is unknown. According to the maximum likelihood criterion, the best estimate of X_v is the value $\tilde{X}_v$ that maximizes the probability density $g(X_A, X_B; X_v)$ (7.28) with respect to X_v. By imposing

$$\frac{\mathrm{d}g(X_A, X_B; X_v)}{\mathrm{d}X_v} = 0, \quad (7.29)$$

and introducing, to simplify notation, the weights

$$w_A = \frac{1}{(\delta X_A)^2} = \frac{1}{\sigma_A^2}, \quad w_B = \frac{1}{(\delta X_B)^2} = \frac{1}{\sigma_B^2}, \quad (7.30)$$

one easily recovers the expression (4.37) of the weighted average

$$\tilde{X}_v = X_0 = \frac{X_A w_A + X_B w_B}{w_A + w_B}. \quad (7.31)$$

The generalization (4.39) to the case of more than two measurements, $X_i \pm \delta X_i$, is immediate.

When the uncertainty of at least one of the measures is not due to random fluctuations and cannot be expressed as the standard deviation of a normal distribution, the previous derivation of the weighted average from the maximum likelihood criterion is, strictly speaking, not valid. The expression of the weighted average can however still be used as a good approximation, provided the uncertainties are expressed as standard deviations of suitable distributions (Sect. 4.5).

Problems

7.1. Toss a coin N times. Plot the relative frequency of the appearance of "head" as a function of the number of tosses.

7.2. Reconsider the logical sequence that led to the expression (4.28) of uncertainty due to random fluctuations in Sect. 4.3. Use the data of Problem 4.3, relative to 200 measures of the period of a pendulum. The table listing the values T_i of the period (top row) and the corresponding number n_i of measures (bottom row) is here repeated.

1.91	1.92	1.93	1.94	1.95	1.96	1.97	1.98	1.99	2.00	2.01	2.02	2.03	2.04	2.05
4	4	2	8	15	19	18	19	56	16	10	19	6	3	1

Calculate the sample mean and estimate the parent mean according to (7.16). Calculate the sample variance and the sample standard deviation. Estimate the parent variance and the parent standard deviation according to the criterion of maximum likelihood (7.23) and compare their values with the values of the unbiased estimator (7.17).

Estimate the mean and variance of the limiting distribution of sample means making use of (7.9) and (7.10).

Part III

Data Analysis

8 Uncertainty in Indirect Measurements

In Chap. 4, the causes of uncertainty in direct measurements have been analyzed, and the standard expression of the result of a measurement, $X_0 \pm \delta X$, has been introduced. The central value X_0 and the uncertainty δX have been expressed in terms of mean value and standard deviation, respectively, of a suitable distribution. In laboratory practice, many physical quantities are, however, indirectly measured, by exploiting functional relations that connect them to other directly measured quantities (Sect. 1.3).

In this chapter, the procedures for propagating the uncertainty of the directly measured quantities to the indirectly measured quantities are presented.

8.1 Introduction to the Problem

Let the quantities $X, Y, Z, \ldots$ be directly measured, whereas the quantity Q is indirectly measured through a functional relation $Q = f(X, Y, Z, \ldots)$; for example, $Q = X + Y$, or $Q = XY/Z$.

Example 8.1. Length ℓ and period T of a pendulum are directly measured. The acceleration of gravity g is indirectly measured through the relation $g = 4\pi^2 \ell / T^2$.

Because the quantities $X, Y, Z, \ldots$ are affected by uncertainty, Q is also affected by uncertainty, and has to be expressed as $Q_0 \pm \delta Q$, say in terms of mean value and standard deviation of a suitable distribution.

The aim of this chapter is to find out how the central value Q_0 can be determined and the uncertainty δQ can be evaluated, from the central values $X_0, Y_0, Z_0, \ldots$ and from the uncertainties $\delta X, \delta Y, \delta Z, \ldots$ of the directly measured quantities, respectively. The uncertainty δQ is sometimes called *combined uncertainty*. The problem is actually rather complex, and it is here gradually solved, starting from the simplest situation, to arrive at an approximate expression of general validity. The reader only interested in final results can skip Sects. 8.2 through 8.4, and go directly to Sect. 8.5.

Statistical Independence of Direct Measurements

A first important point to clarify concerns the statistical independence of the directly measured quantities.

Let us, for simplicity, consider only two quantities, X and Y, directly measured, and suppose that their limiting distributions and their means X_0 and Y_0 are known with good approximation. If now two single values x and y are considered, their deviations from the mean, $x - X_0$ and $y - Y_0$, can be calculated. The measures of X and Y are said to be statistically independent if the deviations $x - X_0$ and $y - Y_0$ are uncorrelated, say if knowledge of the deviation $x - X_0$ gives no information at all on the deviation $y - Y_0$, and viceversa. Otherwise stated, the deviations $x - X_0$ and $y - Y_0$ can be considered as independent random variables (Sect. 6.8). The concept of statistical independence can be generalized to any number of physical quantities.

The statistical independence introduced here has nothing to do with the possible correlation between the values of the physical quantities, that is considered in Chap. 10.

Example 8.2. Length and period of a pendulum are directly measured. It is immediate to check that the two quantities are correlated: when the length increases, the period increases as well. The measures of length and period are, however, statistically independent, because their deviations from the mean are uncorrelated.

Example 8.3. One seeks to measure the perimeter P of a polygon by a meter stick. To this aim, one directly measures the sides $a, b, c, \ldots$ of the polygon, and one calculates the perimeter as a sum: $P = a + b + c + \cdots$. Let us now suppose that the main cause of uncertainty is the lack of confidence on the calibration of the meter stick; one is thus induced to suppose that all measures are similarly biased (in excess or in defect), and their uncertainties are thus not independent.

In Sects. 8.2 and 8.3, the propagation of uncertainty for statistically independent measures is thoroughly treated; the case of nonindependent measures is briefly treated in Sect. 8.4.

8.2 Independent Quantities, Linear Functions

The procedure of propagation of uncertainty is relatively simple when the functional relation between Q and $X, Y, Z, \ldots$ is linear:

$$Q = a + bX + cY + dZ + \cdots \qquad (8.1)$$

where $a, b, c, d, \ldots$ are constant coefficients. The general expression (8.1) includes the important cases

of addition $\qquad Q = X + Y$,
of subtraction $\qquad Q = X - Y$,
of direct proportionality $\quad Q = bX$.

If the relation connecting Q to $X, Y, Z, \ldots$ is linear, the mean and the standard deviation of the distribution of the Q values can be easily obtained from the means and the standard deviations of the distributions of the $X, Y, Z, \ldots$ values, by exploiting some general properties of means and variances, introduced in Sect. 7.2, equations (7.5) through (7.8), and demonstrated in Appendix D.8.

For a linear relation such as (8.1), the mean $\mathbf{m}[Q]$ is connected to the means of $X, Y, Z, \ldots$ by

$$\mathbf{m}[Q] = a + b\,\mathbf{m}[X] + c\,\mathbf{m}[Y] + d\,\mathbf{m}[Z] + \cdots. \qquad (8.2)$$

If the values $X, Y, Z, \ldots$ are statistically independent, the variance $\mathbf{D}[Q]$ is connected to the variances of $X, Y, Z, \ldots$ by

$$\mathbf{D}[Q] = b^2\,\mathbf{D}[X] + c^2\,\mathbf{D}[Y] + d^2\,\mathbf{D}[Z] + \cdots, \qquad (8.3)$$

and the standard deviations are connected by

$$\sigma[Q] = \sqrt{b^2\,\mathbf{D}[X] + c^2\,\mathbf{D}[Y] + d^2\,\mathbf{D}[Z] + \cdots}. \qquad (8.4)$$

The central value of a quantity is expressed by the mean of the distribution, therefore from (8.2) one gets

$$Q_0 = a + b\,X_0 + c\,Y_0 + d\,Z_0 + \cdots. \qquad (8.5)$$

The uncertainty is expressed by the standard deviation of a suitable distribution, therefore from (8.4) one gets

$$\delta Q = \sqrt{b^2\,(\delta X)^2 + c^2\,(\delta Y)^2 + \cdots}. \qquad (8.6)$$

The uncertainty δQ is thus obtained by quadratically summing the uncertainties $\delta X, \delta Y, \ldots$ weighted by the coefficients $b^2, c^2, \ldots$. It is worth noting that the quadratic sum of two numbers is smaller than their direct sum: $(s^2 + t^2)^{1/2} \leq s + t$.

Example 8.4. The weighted average was introduced in Sect. 4.4, and further considered in Sect. 7.3:

$$X_w = \frac{\sum_i X_i w_i}{\sum_i w_i}, \quad \delta X_w = \frac{1}{\sqrt{\sum_i w_i}}, \quad \text{where } w_i = \frac{1}{(\delta X_i)^2}.$$

The expression of the uncertainty δX_w can now be explained. The weighted average is a particular case of the linear function (8.1); X_w corresponds to Q, and the terms X_i are multiplied by the coefficients $w_i / \sum w_i$. Using (8.6) and remembering that $w_i = 1/(\delta X_i)^2$, one gets

8 Uncertainty in Indirect Measurements

$$(\delta X_w)^2 = \frac{1}{(\sum w_i)^2} \sum w_i^2 (\delta X_i)^2 = \frac{1}{(\sum w_i)^2} \sum w_i = \frac{1}{\sum w_i}.$$

Let us now consider the application of (8.5) and (8.6) to some simple cases.

Sum of Quantities: $Q = X + Y$

For an addition, $Q = X + Y$, from (8.5) and (8.6) one gets

$$Q_0 = X_0 + Y_0, \qquad (\delta Q)^2 = (\delta X)^2 + (\delta Y)^2. \qquad (8.7)$$

The generalization of (8.7) to the sum of more than two quantities is immediate.

Example 8.5. To measure the pressure of a gas in a tank, one utilizes two instruments: a manometer to measure the gas pressure relative to the atmospheric pressure, P_{rel}, and a barometer to measure the atmospheric pressure itself, P_{atm}. The two measurements are statistically independent, and their values are: $P_{\text{rel}} = (0.475 \pm 0.004)$ bar and $P_{\text{atm}} = (0.988 \pm 0.002)$ bar. The gas pressure is $P = P_{\text{rel}} + P_{\text{atm}}$. According to (8.7), $P_0 = (0.475 + 0.998) = 1.473$ bar, $\delta P = [(\delta P_{\text{rel}})^2 + (\delta P_{\text{atm}})^2]^{1/2} = 0.0044$ bar.

Example 8.6. In Sect. 4.5, it was suggested that the uncertainties of the same quantity due to different causes should be quadratically summed (equation 4.46). The procedure is supported by (8.7) if the causes of uncertainty are statistically independent. In fact, let us suppose that the result of a measurement is $X_0 \pm \delta X_a$; taking into account a further independent uncertainty δX_b is formally equivalent to summing up the quantity $0 \pm \delta X_b$ to the quantity $X_0 \pm \delta X_a$.

Difference of Quantities: $Q = X - Y$

For a subtraction, $Q = X - Y$, from (8.5) and (8.6) one gets

$$Q_0 = X_0 - Y_0, \qquad (\delta Q)^2 = (\delta X)^2 + (\delta Y)^2. \qquad (8.8)$$

The central values are subtracted, but the uncertainties are quadratically summed, as with additions.

Example 8.7. The mass of an empty calorimeter is $m_c = (257.3 \pm 0.1)$ g. After an amount of water has been poured into the calorimeter, the total mass becomes $m_t = (298.5 \pm 0.1)$ g. The net mass of water can be obtained by difference, $m_0 = m_t - m_c = 298.5 - 257.3 = 41.2$ g. The uncertainty is $\delta m = [(0.1)^2 + (0.1)^2]^{1/2} = 0.14$ g.

Great attention must be paid when subtracting two very similar values X_0 and Y_0: it can happen that the difference is comparable with the uncertainty. For example, $(251 \pm 1) - (250 \pm 1) = 1 \pm 1.4$.

Direct Proportionality: $Q = b\,X$

For a direct proportionality, $Q = bX$, from (8.5) and (8.6) one gets

$$Q_0 = b\,X_0\,, \qquad \delta Q = |b|\,\delta X\,. \tag{8.9}$$

Example 8.8. The wavelength λ and the period T of an electromagnetic wave that propagates in vacuum are connected through the relation $\lambda = c\,T$, where the speed of light c is an exact constant (Appendix C.2). The uncertainty δT of the period propagates to the wavelength as $\delta\lambda = c\,\delta T$.

The coefficient b in (8.9) is the first derivative of Q with respect to X, $b = \mathrm{d}Q/\mathrm{d}X$. For a linear relation, the coefficient b is constant, and measures the slope of the straight line representing the function $Q = bX$ (Fig. 8.1). The propagation of the uncertainty from X to Q thus depends on the slope of the straight line.

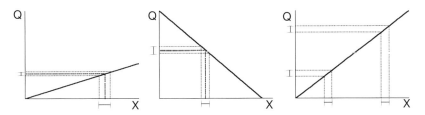

Fig. 8.1. Illustrating the propagation of uncertainty for the direct proportionality $Q = bX$. The uncertainty δQ depends on δX and on the slope b of the straight line: $\delta Q = |b|\,\delta X$ (left and center). The uncertainty δQ is instead independent of the value X_0 (right).

8.3 Independent Quantities, Nonlinear Functions

If Q is connected to $X, Y, Z, \ldots$ by a nonlinear relation, calculating the central value Q_0 and the uncertainty δQ is a rather complex problem, that can only be approximately solved.

Functions of One Variable: $Q = f(X)$

To better understand the origin of the difficulties and the philosophy underlying their solution, it is convenient to begin with the simplest case, say a nonlinear function $Q = f(X)$ of only one variable X. Let us suppose that $X_0 \pm \delta X$ is known through a direct measurement, and let us try to answer the following questions. Is it reasonable to set $Q_0 = f(X_0)$? And how can δQ be calculated from δX? As a working example, let us focus on the simple case $Q = \beta X^2$.

8 Uncertainty in Indirect Measurements

Example 8.9. From the direct measurement of the diameter $2R$, the radius $R_0 \pm \delta R$ of a cylinder is determined. The section S of the cylinder can now be indirectly measured as $S = \pi R^2$. How can the value of the cylinder section be expressed in the standard form $S_0 \pm \delta S$?

The graph of the function $Q = \beta X^2$ is a parabola (Fig. 8.2). Let us consider a generic value X_0 and the uncertainty interval $2\delta X$ centered on X_0. The uncertainty interval $2\delta Q$ corresponding to $2\delta X$ can be graphically determined (dotted lines in Fig. 8.2, left): the value Q_0 at the center of the uncertainty interval $2\delta Q$ does not correspond to βX_0^2. Also, the same uncertainty δX corresponds to different uncertainties δQ when X_0 is varied (Fig. 8.2, center).

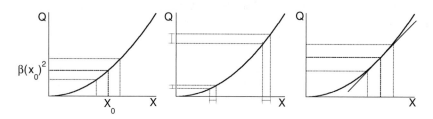

Fig. 8.2. Illustration of the uncertainty propagation for $Q = \beta X^2$. Left: the center Q_0 of the uncertainty interval $2\delta Q$ does not correspond to βX_0^2. Center: when X_0 is changed, different intervals δQ correspond to the same interval δX. Right: local linearization of $Q = \beta X^2$.

Approximation of Local Linearity

The calculation of Q_0 and δQ for a nonlinear function $Q = f(X)$ is much simplified by the approximation of local linearity. The function $f(X)$ is replaced, within the uncertainty interval $2\delta X$, by a straight line $Q = a + bX$, tangent to the curve $f(X)$ for $X = X_0$ (Fig. 8.2, right):

$$Q = f(X) \quad \rightarrow \quad Q = a + bX . \tag{8.10}$$

Thanks to this approach, one can extend the procedure developed for direct proportionality (8.9) also to the case of nonlinearity. The difference is that the procedure is now approximate.

The central value Q_0 is approximated as $f(X_0)$, and the uncertainty δQ is approximated by $|b|\,\delta X$, where now b is the first derivative of Q with respect to X, calculated for $X = X_0$:

$$Q_0 \simeq f(X_0) , \qquad \delta Q \simeq \left|\frac{dQ}{dX}\right|_0 \delta X . \tag{8.11}$$

8.3 Independent Quantities, Nonlinear Functions

Example 8.10. Let us reconsider Example 8.9, concerning the measurement of the radius and section of a cylinder, and suppose that the radii of two different cylinders have been measured, with the same uncertainty $\delta R = 0.01$ mm. The radius of the first cylinder is $R_1 = (0.5 \pm 0.01)$ mm and the section, according to (8.11), is $S_1 = \pi R_1^2 = (0.78 \pm 0.03)$ mm². The radius of the second cylinder is $R_2 = (5 \pm 0.01)$ mm and the section is $S_2 = \pi R_2^2 = (78.5 \pm 0.3)$ mm². In spite of the uncertainties of the radii being the same, the absolute uncertainty of the section is ten times larger for the second cylinder than for the first one. Conversely, the relative uncertainty is ten times smaller for the first cylinder.

One can easily understand that, the smaller is the uncertainty δX with respect to the central value X_0, the better is the approximation of local linearity.

Raising to Power: $Q = X^n$

A particularly important case of relation $Q = f(x)$ is the raising to power, $Q = X^n$. Using (8.11) one finds

$$Q_0 \simeq X_0^n, \quad \delta Q \simeq n \left| X_0^{n-1} \right| \delta X \quad \text{say} \quad \frac{\delta Q}{|Q_0|} \simeq n \frac{\delta X}{|X_0|} . \tag{8.12}$$

The relative uncertainty of Q is n times larger than the relative uncertainty of X.

General Expression

The treatment introduced for a function of one variable, $Q = f(X)$, can now be extended to the general case $Q = f(X, Y, Z, \ldots)$. To simplify notation without loss of generality, it is sufficient to consider a function of two variables, $Q = f(X, Y)$ (typical examples are $Q = XY$ and $Q = X/Y$).

The approximate calculation of Q_0 and δQ is again based on the local linearization of the function $f(X, Y)$ in a surrounding of the central values (X_0, Y_0), so that the actual surface $f(X, Y)$ is substituted by the plane tangent to the surface in correspondence of (X_0, Y_0):

$$Q = f(X, Y) \quad \rightarrow \quad Q = a + bX + cY . \tag{8.13}$$

The coefficients b and c in (8.13) correspond to the first partial derivatives of Q with respect to X and Y, calculated for $X = X_0$ and $Y = Y_0$, respectively:

$$b = \left(\frac{\partial Q}{\partial X} \right)_0, \quad c = \left(\frac{\partial Q}{\partial Y} \right)_0 . \tag{8.14}$$

It is worth remembering here that the first partial derivative of a function $Q = f(X, Y, Z, \ldots)$ with respect to one of the independent variables, for example,

X, is calculated as a normal derivative, considering the other variables as constant parameters.

The problem is again reduced, although locally and approximately, to the linear case of Sect. 8.2. If the quantities X and Y are statistically independent, then

$$Q_0 \simeq f(X_0, Y_0), \quad (\delta Q)^2 \simeq \left(\frac{\partial Q}{\partial X}\right)_0^2 (\delta X)^2 + \left(\frac{\partial Q}{\partial Y}\right)_0^2 (\delta Y)^2. \tag{8.15}$$

The uncertainty δQ is obtained from the quadratic sum of the uncertainties δX and δY, weighted by the corresponding partial derivatives.

The generalization to $Q = f(X, Y, Z, \ldots)$ is immediate:

$$(\delta Q)^2 \simeq \left(\frac{\partial Q}{\partial X}\right)_0^2 (\delta X)^2 + \left(\frac{\partial Q}{\partial Y}\right)_0^2 (\delta Y)^2 + \left(\frac{\partial Q}{\partial Z}\right)_0^2 (\delta Z)^2 + \cdots \tag{8.16}$$

Again, the smaller are the uncertainties $\delta X, \delta Y, \ldots$ with respect to the central values, the better is the approximation of local linearity.

Let us now consider the application of (8.15) to some simple cases.

Product: $Q = XY$

For a product $Q = XY$, the central value is approximated by

$$Q_0 \simeq X_0 Y_0. \tag{8.17}$$

The uncertainty of the product, according to (8.15), is

$$(\delta Q)^2 \simeq \left(\frac{\partial Q}{\partial X}\right)_0^2 (\delta X)^2 + \left(\frac{\partial Q}{\partial Y}\right)_0^2 (\delta Y)^2 = Y_0^2 (\delta X)^2 + X_0^2 (\delta Y)^2. \tag{8.18}$$

Equation (8.18) can be recast in a simpler form if both members are divided by $Q_0^2 = X_0^2 Y_0^2$:

$$\left(\frac{\delta Q}{Q_0}\right)^2 \simeq \left(\frac{\delta X}{X_0}\right)^2 + \left(\frac{\delta Y}{Y_0}\right)^2. \tag{8.19}$$

The *relative uncertainty* of the product Q is the quadratic sum of the relative uncertainties of the factors X and Y.

Example 8.11. The two sides of a rectangle are directly measured; their values are $a = 13 \pm 0.05$ cm and $b = 25 \pm 0.05$ cm, so that the relative uncertainties are $\delta a/a_0 = 0.0038$ and $\delta b/b_0 = 0.002$, respectively. The area of the rectangle is now calculated as $S = ab$. The central value is $S_0 = a_0 b_0 = 325$ cm^2; to evaluate the uncertainty, one first calculates $(\delta S/S_0)^2 = (\delta a/a_0)^2 + (\delta b/b_0)^2 = 1.8 \times 10^{-5}$, then one recovers $\delta S/S_0 = 0.0043$, and finally $S = 325 \pm 1.4$ cm^2.

Quotient: $Q = X/Y$

For a quotient $Q = X/Y$, the central value is approximated by

$$Q_0 \simeq X_0/Y_0 \, . \tag{8.20}$$

The uncertainty of the quotient, according to (8.15), is

$$(\delta Q)^2 \simeq \left(\frac{\partial Q}{\partial X}\right)_0^2 (\delta X)^2 + \left(\frac{\partial Q}{\partial Y}\right)_0^2 (\delta Y)^2 = \frac{1}{Y_0^2}(\delta X)^2 + \frac{X_0^2}{Y_0^4}(\delta Y)^2 \, . \tag{8.21}$$

Equation (8.21) can be recast in a simpler form if both members are divided by $Q_0^2 = X_0^2/Y_0^2$:

$$\left(\frac{\delta Q}{Q_0}\right)^2 \simeq \left(\frac{\delta X}{X_0}\right)^2 + \left(\frac{\delta Y}{Y_0}\right)^2 \, . \tag{8.22}$$

As with the product, also with a quotient, the *relative uncertainty* of Q is the quadratic sum of the relative uncertainties of X and Y.

Example 8.12. The difference of electric potential across a resistor and the electric current are directly measured; their values are $V = 200 \pm 0.1$ V and $I = 50 \pm 0.1$ mA, respectively, so that the relative uncertainties are $\delta V/V_0 = 0.0005$ and $\delta I/I_0 = 0.002$, respectively. The value of the resistance R can now be obtained as $R = V/I$. The central value is $R_0 = V_0/I_0 = 4$ kΩ; to evaluate the uncertainty, one first calculates $(\delta R/R_0)^2 = (\delta V/V_0)^2 + (\delta I/I_0)^2 = 4.2 \times 10^{-6}$, then one recovers $\delta R/R_0 = 2.1 \times 10^{-3}$, and finally $R = 4 \pm 0.008$ kΩ.

8.4 Nonindependent Quantities

Up to now, the propagation of uncertainty has been studied for statistically independent measures. The statistical independence is, however, not always guaranteed. Two simple examples help in understanding the difference between statistically independent and nonindependent measures.

Example 8.13. To determine the perimeter P of a square, one directly measures its side, $a_0 \pm \delta a$. Let us now compare two different procedures for calculating the perimeter P and its uncertainty. By the first procedure, the perimeter is calculated as $P = 4a$, and the uncertainty, according to (8.9), is $\delta P = 4\,\delta a$. By the second procedure, the perimeter is calculated as $P = a + a + a + a$; if the uncertainty is evaluated according to (8.7), one obtains $\delta P = 2\,\delta a$. This second evaluation of uncertainty is, however, wrong, because the four added quantities are identical, and then not statistically independent, so that the use of (8.7) is incorrect.

Example 8.14. To determine the area S of a square, one directly measures its side, $a_0 \pm \delta a$. Let us again compare two different procedures for calculating the area S and its uncertainty. By the first procedure, the area is calculated as $S = a^2$, and the uncertainty, according to (8.11), is $\delta S = 2a\,\delta a$. By the second procedure, the area is calculated as $S = aa$; if the uncertainty is evaluated according to (8.18), one obtains $\delta S = \sqrt{2}a\,\delta a$. This second evaluation of uncertainty is again wrong, because the two factors are identical, and then not statistically independent, so that the use of (8.18) is incorrect.

The two examples show that the propagation rules based on quadratic sums are wrong if the direct measurements are not statistically independent. In the following, a deeper understanding will be gained by considering the particular case $Q = Q(X,Y)$, where the two quantities X and Y are directly and contemporarily measured, and their uncertainty is obtained from an equal number of repeated measurements.

Repeated Direct Measurement of Two Quantities X and Y

Let us repeat N times the direct measurement of the two quantities X and Y. The result of each measurement is a pair of values (x_i, y_i). The result of the full set of measurements is a couple of sample distributions: one for the x values, with mean m_x^* and variance D_x^*, the other for the y values, with mean m_y^* and variance D_y^*.

According to the convention of Sect. 4.3, the central values of the two quantities are estimated by their sample means:

$$X_0 = m_x^*, \quad Y_0 = m_y^*, \qquad (8.23)$$

and the uncertainties are estimated by the standard deviations of the distributions of sample means:

$$\delta X = \tilde{\sigma}[m_x^*] = \sqrt{\frac{D_x^*}{N-1}}, \quad \delta Y = \tilde{\sigma}[m_y^*] = \sqrt{\frac{D_y^*}{N-1}}. \qquad (8.24)$$

Let us now consider the indirect measurement of $Q(X,Y)$. To each pair of values (x_i, y_i) corresponds a value of Q: $q_i = f(x_i, y_i)$. One can then build up a sample distribution of q_i values, with mean m_q^* and variance D_q^*, so that the central value and the uncertainty of Q are

$$Q_0 = m_q^*, \qquad \delta Q = \sqrt{\frac{D_q^*}{(N-1)}}. \qquad (8.25)$$

Our goal is now to find the functional relations between the central value Q_0 and the central values X_0, Y_0, and between the uncertainty δQ and the uncertainties $\delta X, \delta Y$, respectively. To simplify notation, we work on variances D_x^*, D_y^*, D_q^* instead of uncertainties $\delta X, \delta Y, \delta Q$.

Taylor Expansions

The procedure is based on the Taylor expansion. For a function $Q = f(X)$ of one variable X, the generic value $q_i = f(x_i)$ can be expressed as a Taylor expansion around a given value $q_0 = f(x_0)$:

$$q_i = f(x_i) = \underbrace{f(x_0) + \left(\frac{dQ}{dX}\right)_0 (x_i - x_0)}_{\text{linear terms}} + \frac{1}{2}\left(\frac{dQ}{dX}\right)_0^2 (x_i - x_0)^2 + \cdots . \quad (8.26)$$

The index 0 means that the derivatives are calculated for $x = x_0$, and the dots $\cdots$ stay for the third- and higher-order terms. The first two terms in (8.26) represent the linear approximation.

For a function $Q = f(X, Y)$ of two variables X, Y, the Taylor expansion is

$$q_i = f(x_i, y_i) = \underbrace{f(x_0, y_0) + \left(\frac{\partial Q}{\partial X}\right)_0 (x_i - x_0) + \left(\frac{\partial Q}{\partial Y}\right)_0 (y_i - y_0)}_{\text{linear terms}} + \cdots . \quad (8.27)$$

Mean, Variance, and Covariance

Let us now evaluate mean and variance of the distribution of the q_i values, by the Taylor expansion (8.27). As a first step, let us substitute m_x^*, m_y^* to x_0, y_0. The expansion (8.27) can be truncated at the first-order term if the distributions of the x_i and y_i values are sufficiently narrow. This corresponds to the local linear approximation of Sect. 8.3.

If the generic value q_i is expressed by (8.27), the central value Q_0 can be calculated as

$$\begin{aligned} Q_0 = m_q^* &= \frac{1}{N}\sum_i q_i \\ &\simeq \frac{1}{N}\sum_i \left[Q(m_x^*, m_y^*) + \left(\frac{\partial Q}{\partial X}\right)_0 (x_i - m_x^*) + \left(\frac{\partial Q}{\partial Y}\right)_0 (y_i - m_y^*)\right] \\ &= Q(m_x^*, m_y^*), \quad (8.28) \end{aligned}$$

which corresponds to the expression $Q_0 \simeq f(X_0, Y_0)$ of (8.15). In the last equality of (8.28), one has taken into account that $\sum(x_i - m_x^*) = \sum(y_i - m_y^*) = 0$.

The sample variance D_q^* can now be evaluated by using the expression of m_q^* calculated in (8.28):

$$D_q^* = \frac{1}{N}\sum_i (q_i - m_q^*)^2$$

$$\simeq \frac{1}{N} \sum_i \left[\left(\frac{\partial Q}{\partial X}\right)_0 (x_i - m_x^*) + \left(\frac{\partial Q}{\partial Y}\right)_0 (y_i - m_y^*) \right]^2$$
$$= \left(\frac{\partial Q}{\partial X}\right)_0^2 D_x^* + 2 \left(\frac{\partial Q}{\partial X}\right)_0 \left(\frac{\partial Q}{\partial Y}\right)_0 \sigma_{xy}^* + \left(\frac{\partial Q}{\partial Y}\right)_0^2 D_y^*, \qquad (8.29)$$

where

$$D_x^* = \frac{1}{N} \sum_i (x_i - m_x^*)^2 \quad \text{and} \quad D_y^* = \frac{1}{N} \sum_i (y_i - m_y^*)^2 \qquad (8.30)$$

are the sample variances of X and Y. The quantity σ_{xy}^* appearing in the last member of (8.29) is the sample covariance of X and Y:

$$\sigma_{xy}^* = \frac{1}{N} \sum_i (x_i - m_x^*)(y_i - m_y^*). \qquad (8.31)$$

In general, the covariance depends on the degree of correlation of two random variables; if the random variables are independent, the covariance of their parent distributions is zero (Sect. 6.8). Here, the sample covariance measures the degree of correlation between the deviations of the x_i and y_i values from their respective means.

Statistically Independent Quantities

If the deviations of the x_i and y_i values from their respective means are completely independent, the parent covariance is zero. As for the sample covariance of (8.29), one expects that $\sigma_{xy}^* \to 0$ for $N \to \infty$. The uncertainty is thus given by

$$(\delta Q)^2 = \left(\frac{\partial Q}{\partial X}\right)_0^2 (\delta X)^2 + \left(\frac{\partial Q}{\partial Y}\right)_0^2 (\delta Y)^2, \qquad (8.32)$$

and (8.15) is recovered.

Statistically Nonindependent Quantities

If the deviations of the x_i and y_i values from their respective means are not independent, the sample covariance σ_{xy}^* in (8.29) is not zero. The sample covariance σ_{xy}^* can be positive or negative, but its value is always connected to the values of the variances D_x^* and D_y^* by the Schwartz inequality:

$$(\sigma_{xy}^*)^2 \leq D_x^* D_y^*, \quad \text{say} \quad |\sigma_{xy}^*| \leq \sqrt{D_x^* D_y^*}. \qquad (8.33)$$

The Schwartz inequality can be demonstrated by considering a function $A(t)$ of a generic variable t, defined as

$$A(t) = (1/N) \sum_i \left[(x_i - m_x^*) + t(y_i - m_y^*)\right]^2$$
$$= D_x^* + 2t\sigma_{xy}^* + t^2 D_y^*. \tag{8.34}$$

For every value of t, one has $A(t) \geq 0$. The Schwartz inequality is demonstrated by substituting $t = -\sigma_{xy}^*/D_y^*$.

As a consequence of the Schwartz inequality, from (8.29) one obtains an expression for the *maximum uncertainty* of the quantity Q,

$$(\delta Q) \leq \left|\frac{\partial Q}{\partial X}\right|_0 \delta X + \left|\frac{\partial Q}{\partial Y}\right|_0 \delta Y, \tag{8.35}$$

that can be easily extended to the general case $Q = f(X, Y, Z, \ldots)$.

8.5 Summary

Let us now summarize the main results obtained in this chapter. It is worth remembering that all conclusions of this chapter are strictly valid only if the uncertainties are expressed as standard deviations of suitable distributions.

If the direct measurements are *statistically independent*, the uncertainty on the indirectly measured value $Q(X, Y, Z, \ldots)$ can be calculated through the general formula (8.16):

$$\delta Q \simeq \sqrt{\left(\frac{\partial Q}{\partial X}\right)_0^2 (\delta X)^2 + \left(\frac{\partial Q}{\partial Y}\right)_0^2 (\delta Y)^2 + \left(\frac{\partial Q}{\partial Z}\right)_0^2 (\delta Z)^2 + \cdots}. \tag{8.36}$$

- If the function $Q(X, Y, Z, \ldots)$ is linear, say $Q = a + bX + cY + dZ + \cdots$, then the equality in (8.36) is exact ($\simeq$ is substituted by $=$).
- In particular, for sums $Q = X + Y$ and differences $Q = X - Y$, the absolute uncertainty δQ is the quadratic sum of the absolute uncertainties δX and δY; see (8.7) and (8.8).
- If the function $Q(X, Y, X, \ldots)$ is not linear, then the equality in (8.36) is only approximate; the smaller are the uncertainties $\delta X, \delta Y, \ldots$ with respect to the central values $X_0, Y_0, \ldots$, the better is the approximation.
- In particular, for products and quotients, the relative uncertainty $\delta Q/Q_0$ is the quadratic sum of the relative uncertainties $\delta X/X_0$ and $\delta Y/Y_0$; see (8.19) and (8.22).
- For a raising to power $Q = X^n$, the relative uncertainty $\delta Q/Q_0$ is n times larger than the relative uncertainty $\delta X/X_0$; see (8.12).

If the direct measurements are *statistically nonindependent*, then (8.36) is not valid. One can in any case demonstrate that the maximum value of the uncertainty of Q is

$$(\delta Q)_{\max} \simeq \left|\frac{\partial Q}{\partial X}\right|_0 \delta X + \left|\frac{\partial Q}{\partial Y}\right|_0 \delta Y + \left|\frac{\partial Q}{\partial Z}\right|_0 \delta Z + \cdots. \tag{8.37}$$

Problems

8.1. The temperature of a room is measured by a thermometer with resolution $\Delta T = 1°C$. The maximum and minimum temperatures measured during the day are $T_{\max} = 23°C$ and $T_{\min} = 20°C$, respectively. Evaluate the standard uncertainties and the relative uncertainties of $T_{\max}$ and of the temperature variation $T_{\max} - T_{\min}$. [*Answer:* 0.3°C, 1.5%, 0.4°C, 13%.]

8.2. A Geiger counter is exposed to a radioactive source during a time interval of 10 minutes; the counts are $n_{\text{tot}} = 25$. The source is then shielded, and the counts, due to the environmental background, are $n_{\text{bgr}} = 12$ in 10 minutes. The number of counts due to the radioactive source is evaluated as $n_{\text{sig}} = n_{\text{tot}} - n_{\text{bgr}} = 13$. Evaluate the absolute and relative uncertainties of n_{tot} and n_{sig}. [*Answer:* 5, 20%, 6, 46%.]

8.3. In a calorimeter, the heat capacity C of a body is indirectly measured as the ratio $Q/\Delta T$, where Q is the heat absorbed from an electric resistor and ΔT is the temperature variation. The heat absorbed is $Q = 2010 \pm 4\,\text{J}$ and the temperature variation is $\Delta T = 4 \pm 0.1\,\text{K}$. Calculate the heat capacity and evaluate its standard uncertainty. [*Answ.:* $C = 502 \pm 13\,\text{J/K}$.]
To reduce the uncertainty of the heat capacity, is it more convenient to increase the accuracy of the measure of heat or of the measure of temperature?

8.4. Length and period of a pendulum are directly measured. Their values are $\ell = 81 \pm 0.3\,\text{cm}$ and $T = 1.8 \pm 0.004\,\text{s}$, respectively. The acceleration of gravity is indirectly measured as $g = (2\pi/T)^2\,\ell$. Evaluate the uncertainty δg. [*Answer:* $\delta g = 0.057\,\text{m/s}^2$.]

9 Confidence Levels

It was shown in Sect. 4.3 that a measure affected by random fluctuations can be expressed as $X = X_0 \pm \delta X$, where the best estimate of the central value X_0 is the sample mean, and the uncertainty is evaluated by the standard deviation of the distribution of sample means.

The standard deviation of the distribution of sample means can, however, only be estimated from a finite number of measurements. The probabilistic interpretation of the estimated uncertainty interval depends on the sample size, as is clarified in this chapter.

9.1 Probability and Confidence

The interpretation of the uncertainty interval is based on the distinction between probability and confidence.

Parent and Sample Populations of Measurement Values

As observed in Chap. 7, the values obtained from N repeated measurements of a physical quantity can be considered as a finite sample of a hypothetical infinite parent population of single measures, whose distribution can generally be approximated by a normal distribution, with mean m and standard deviation σ. If the parent population were known, its mean m would be the *true value* X_v of the physical quantity.

Actually, only the sample of N values is known, from which one can calculate the sample mean m^*. The sample mean is a random variable, and can in turn be considered as an individual of the infinite parent population of the sample means (relative to a sample of N measures). It was shown in Sect. 7.2 that the distribution of sample means has the following properties.

(a) It is centered at the true value, $\mathbf{m}[m^*] = m = X_v$.
(b) It can be considered normal with very good approximation, if N is sufficiently large.
(c) Its standard deviation $\sigma[m^*]$ is connected to the standard deviation σ of the parent distribution of single measures by $\sigma[m^*] = \sigma/\sqrt{N}$.

Probability Interval

Let us begin our discussion with the abstract hypothesis that the parent population of single measures is completely known. In this case, the true value of the quantity would be exactly known as well: $X_v = m = \mathbf{m}[m^*]$. Moreover, one could answer the following question. If a sample of N measures is available, what is the probability that the sample mean m^* is included in a given interval centered at the true value $X_v = m = \mathbf{m}[m^*]$? It is convenient to measure the different possible intervals taking as the unit the standard deviation of the distribution of sample means $\sigma[m^*]$ (Fig. 9.1, left). An interval is thus defined by the relation

$$m - k\sigma[m^*] < m^* < m + k\sigma[m^*], \tag{9.1}$$

equivalent to

$$|m^* - m| < k\sigma[m^*], \tag{9.2}$$

or to

$$\frac{|m^* - m|}{\sigma[m^*]} < k. \tag{9.3}$$

The parameter k is called the *coverage factor*. Because the distribution of sample means m^* is, to a good approximation, normal, it is convenient to refer to the corresponding standard variable z (Sect. 6.5),

$$z = \frac{m^* - m}{\sigma[m^*]}, \tag{9.4}$$

and rewrite (9.3) as

$$|z| < k. \tag{9.5}$$

The probability $\mathcal{P}_k$ that the sample mean m^* satisfies (9.1), for a given value k of the coverage factor, can be evaluated by the integral of the normal standard distribution:

$$\mathcal{P}_k = \mathcal{P}\{|z| < k\} = \frac{1}{\sqrt{2\pi}} \int_{-k}^{+k} \exp[-z^2/2] \, dz. \tag{9.6}$$

For example, if $k = 1$, the probability is $\mathcal{P}_{k=1} = 68.27\%$, calculated using the tables of Appendix C.3. Conversely, if a value $\mathcal{P}$ of probability is given, one can determine the value of k that satisfies (9.6), again using the tables of Appendix C.3; for example, the probability $\mathcal{P} = 90\%$ is obtained for a coverage factor $k = 1.64$.

It is worth remembering that, in the abstract case considered here, the values $X_v = m$ and $\sigma[m^*]$ are supposed to be known, whereas the random variable m^* is unknown. Equation (9.1) is a direct application of probability theory, and the interval included between $m - k\sigma[m^*]$ and $m + k\sigma[m^*]$ is a *probability interval*.

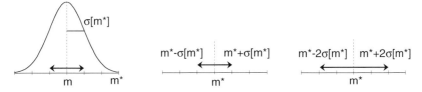

Fig. 9.1. Left: distribution of sample means m^*; the arrow $\leftrightarrow$ shows a probability interval centered at the parent value m, of half-width $\sigma[m^*]$. Center and right: confidence intervals centered on the experimental sample mean m^*, of half-width $\sigma[m^*]$ and $2\sigma[m^*]$, respectively.

Confidence Interval

Let us now consider a realistic case. A physical quantity has been measured N times, and the sample mean m^* has been calculated. The situation is now reversed with respect to the previous abstract case: the fixed value $X_v = m$ is now unknown, whereas a value m^* of a random variable is known.

It was shown, in Sect. 7.3, that the sample mean is a consistent and unbiased estimator of the parent mean m:

$$\tilde{m} = m^* = \frac{1}{N} \sum_i x_i \,. \tag{9.7}$$

The estimate (9.7) is a *point estimate*, represented by a single numerical value m^*. A point estimate of the true value is, by itself, of little interest, inasmuch as it is a random variable, depending on the sample.

A more significant estimate should suitably take into account the degree of randomness of the value $\tilde{m} = m^*$. To this aim, it is necessary to answer the following question. What is the probability that an interval centered on the sample mean m^*, and of given width, contains the true value $X_v = m$? Because $X_v = m$ is not a random variable, but a fixed (although unknown) value, it would be incorrect to refer to this interval as a probability interval. For that reason, it is called the *confidence interval*, and the corresponding probability is called the *coverage probability* or *confidence level*. The confidence interval is schematically represented in Fig. 9.1 (center and right).

To proceed step by step, let us again make an abstract hypothesis, say that, in spite of $X_v = m$ being unknown, the standard deviation σ of the parent population is known. In this case $\sigma[m^*]$ is also known, and a confidence interval can be defined by the relation

$$m^* - k\,\sigma[m^*] \;<\; m \;<\; m^* + k\,\sigma[m^*]\,, \tag{9.8}$$

which is analytically equivalent to (9.2):

$$|\tilde{m} - m| \;<\; k\,\sigma[m^*]\,. \tag{9.9}$$

The coverage factor k of (9.8) can again be connected to the corresponding standard normal variable (9.4).

The standard deviation $\sigma[m^*]$ of the distribution of sample means is conventionally assumed as uncertainty δX of the measure. If $\sigma[m^*]$ is known, the confidence level can again be evaluated by (9.6). For example, the probability $\mathcal{P}_{k=1}$ that the confidence interval $\pm\sigma[m^*]$ contains the true value $X_v = m$ is 68.27%.

In real cases, the standard deviation $\sigma[m^*]$ is also unknown, like the true value $X_v = m$. The value of $\sigma[m^*]$ can only be estimated (Sects. 4.3 and 7.2):

$$\tilde{\sigma}[m^*] = \frac{\tilde{\sigma}}{\sqrt{N}} = \frac{\sigma^*}{\sqrt{N-1}} = \sqrt{\frac{1}{N(N-1)} \sum (x_i - m^*)^2} \,. \qquad (9.10)$$

The estimate of $\sigma[m^*]$ (9.10) has a random character, because it is calculated from a particular sample: the smaller is the number N of measurements, the larger is the degree of randomness of the estimate. The expression (9.8) of the confidence interval is now substituted by

$$m^* - k\tilde{\sigma}[m^*] < m < m^* + k\tilde{\sigma}[m^*] \,, \qquad (9.11)$$

equivalent to

$$\frac{|m^* - m|}{\tilde{\sigma}[m^*]} < k \,. \qquad (9.12)$$

It is convenient to introduce a new random variable

$$t = \frac{m^* - m}{\tilde{\sigma}[m^*]} \,, \qquad (9.13)$$

and rewrite (9.12) as

$$|t| < k \,. \qquad (9.14)$$

The random variable t is different from the standard normal variable (9.4), because $\tilde{\sigma}[m^*]$ is only a random estimate of $\sigma[m^*]$.

One is interested in evaluating the probability (confidence level) that the true value $X_v = m$ satisfies (9.11), for a given value k of the coverage factor, say in evaluating the probability

$$\mathcal{P}'_k = \mathcal{P}\{|t| < k\} \,. \qquad (9.15)$$

The prime in (9.15) is used to distinguish the confidence level $\mathcal{P}'_k$ from the probability $\mathcal{P}_k$ of (9.6).

Approximate Evaluation of the Confidence Level

The confidence level (9.15) can still be approximately evaluated through (9.6), by substituting $\sigma[m^*]$ with the estimate $\tilde{\sigma}[m^*]$. This approximation consists of substituting the variable t of (9.13) with the standard normal variable z of (9.4). This substitution gives probability values approximated in excess, as shown below. The larger is the number N of measurements, the better is expected to be the approximation. In fact, when N increases, the distribution of sample means m^* approaches the normal shape, and the dispersion of the estimates $\tilde{\sigma}[m^*]$ with respect to $\sigma[m^*]$ decreases, so that the value of the variable t approaches the value of z.

9.2 The Student Distribution

A nonapproximate evaluation of the confidence level from a finite sample of N measures can be attempted only if the shape of the parent distribution is known. If the parent distribution is normal, the distribution of the random variable t, defined in (9.13), can be calculated; it is called the *Student distribution* (after the pseudonym of W. S. Gosset, who first introduced it in 1908).

Because, according to (9.10), the estimate $\tilde{\sigma}[m^*]$ depends on the number N of measures, the Student distribution is expected to depend on N as well. Actually, to calculate $\tilde{\sigma}[m^*]$, it is necessary to previously know m^*, which is in turn calculated from the N measures by (9.7). Only $N-1$ values x_i in (9.10) are thus independent, because, once $N-1$ values are known, the Nth value is determined by (9.7). The problem has $\nu = N-1$ degrees of freedom.

The Student distribution depends on the number ν of degrees of freedom. Two Student distributions $S_\nu(t)$ for two different degrees of freedom, $\nu = 1$ and $\nu = 3$, are plotted in Fig. 9.2 (left), where the standard normal distribution is also shown for comparison. One can demonstrate that the Student distribution for $\nu = 1$ corresponds to the Cauchy–Lorentz distribution (Sect. 6.7). When ν increases, say when the number of measures N increases, the shape of the distribution is progressively modified, and for $\nu \to \infty$ the Student distribution tends to the standard normal distribution.

The integrals of the Student distribution are tabulated in Appendix C.4 for different values of ν.

From the Student distribution, one can calculate the confidence level for a given confidence interval, say for a given value k of the coverage factor:

$$\mathcal{P}'_k = \mathcal{P}\{|t| < k\} = \int_{-k}^{+k} S_\nu(t)\, dt . \qquad (9.16)$$

For example, for the standard uncertainty $\delta X = \tilde{\sigma}[m^*]$, say for the coverage factor $k = 1$, the probability is the integral of the Student distribution from $t = -k = -1$ to $t = +k = +1$. The values, calculated using the tables of Appendix C.4, are shown in Fig. 9.2 (right) as a function of the number of degrees of freedom $\nu = N - 1$. The values of the confidence level $\mathcal{P}'_{k=1}$ are smaller than the probability 68.27% corresponding to the standard normal distribution, but approach it for $\nu \to \infty$.

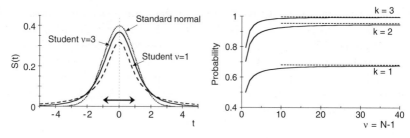

Fig. 9.2. Left: Student distributions for $\nu = 1$ and $\nu = 3$ (dashed and continuous lines, respectively), compared with the standard normal distribution (dotted line); The arrow ↔ shows the confidence interval for the coverage factor $k = 1$. Right: confidence levels as a function of the degrees of freedom, calculated by integrating the corresponding Student distributions, for coverage factors $k = 1, 2$, and 3, respectively; the dashed horizontal straight lines show the probabilities for the standard normal distribution.

9.3 Applications of the Confidence Level

Let us now consider some important applications of the confidence level to the procedures of data analysis.

Uncertainty and Coverage Factor

The conventional expression of the standard uncertainty $\delta X = \tilde{\sigma}[m^*]$ corresponds to a coverage factor $k = 1$. The corresponding confidence level depends on the number N of measurements, say on the number $\nu = N - 1$ of degrees of freedom (Fig. 9.2, right), and asymptotically tends to 68.26% when $N \to \infty$. To guarantee a correct interpretation of the value of a physical quantity, expressed as $X_0 \pm \delta X$, one should always explicitly also give the number of degrees of freedom $\nu = N - 1$, say the size of the sample.

In some applications, it can be convenient to use coverage factors larger than $k = 1$; typical examples are measurements concerning safety where confidence approaching 100% is desirable. To coverage factors larger than one correspond larger confidence intervals, and one speaks of *expanded uncertainty*. The use of coverage factors larger than one is unconventional, and should always be accompanied by explicit warnings. The confidence levels corresponding to coverage factors $k = 2$ and $k = 3$ are shown in Fig. 9.2 (right) as a function of the number of degrees of freedom.

Frequently, when expanded uncertainties are used, one prefers to establish in advance the sought confidence level (such as $\mathcal{P}' = 90\%$, or $\mathcal{P}' = 95\%$), and then determine the corresponding coverage factor k as a function of the number ν of degrees of freedom (see Table C.6 of Appendix C.4).

9.3 Applications of the Confidence Level

Comparison of Different Measurements of the Same Quantity

The concept of confidence intervals allows one to estimate, on probabilistic grounds, the degree of agreement between the results of two different measurements of the same physical quantity. The application is better illustrated by an example.

Example 9.1. The measurement of a given quantity, repeated $N = 20$ times, has led to the following result (in arbitrary units): $X_0 = m^* = 2.5$; $\delta X = \tilde{\sigma}[m^*] = 0.5$. Is this result consistent with the value $m = 1.8$ quoted in the current literature and assumed as the true value? The relative discrepancy is $t = (m^* - m)/\tilde{\sigma}[m^*] = (2.5 - 1.8)/0.5 = 1.4$. Because N is quite large, it is reasonable to approximate t by z. According to the tables of the standard normal distribution (Appendix C.3), the probability that $|m^* - m| < 1.4\,\sigma[m^*]$ because of statistical fluctuations is about 84%. Our experimental result $m^* = 2.5$ is then considered consistent with the "true" value $m = 1.8$, if a 16% probability of finding by chance relative discrepancies larger than $t = 1.4$ is considered reasonable.

Rejection of Data

In a sample of N values x_i (such as N measures of a physical quantity), it can happen that one of the values, let it be x_s, appears to be inconsistent with the other $N-1$ values. To focus our attention on a concrete case, let us suppose that the results of $N = 6$ measurements of a physical quantity are (in arbitrary units)

$$2.7 \quad 2.5 \quad 2.8 \quad 1.5 \quad 2.4 \quad 2.9 \,. \tag{9.17}$$

The value $x_s = 1.5$ appears to be inconsistent with the other values. The possible causes of discrepancy are

- A particularly large statistical fluctuation within the parent population
- An unidentified systematic error
- A new unexpected physical effect

Let us suppose that the last possibility can be excluded. We want to evaluate, on probabilistic grounds, whether the discrepancy is due to a statistical fluctuation or to a systematic error. In the first case (statistical fluctuation) the rejection of x_s would be unjustified. In the second case (systematic error) it could be reasonable to eliminate the value x_s from the sample.

The six values (9.17) can be considered as a sample of a parent distribution of single measures whose mean m and standard deviation σ can be estimated as

$$\tilde{m} = m^* = 2.46\,, \qquad \tilde{\sigma} = \sqrt{\frac{N}{N-1}}\,\sigma^* = 0.5\,. \tag{9.18}$$

The relative discrepancy between x_s and $\tilde{m}$ is

$$t = \frac{x_s - \tilde{m}}{\tilde{\sigma}} = 1.92 \simeq 2 . \tag{9.19}$$

According to Table C.5 of Appendix C.4, taking into account that the problem has $\nu = 5$ degrees of freedom, the probability that t is by chance larger than 2 is about 10%. Otherwise stated, out of $N = 6$ measurements, we expect on the average 0.6 values that differ more than 2σ from the mean, say that are "worse" than x_s.

A conventional criterion, the so-called *Chauvenet criterion*, consists of considering that a value x_s can be rejected from a given sample if the expected number of "worse" values is smaller than 0.5. In this situation, one considers that the discrepancy of x_s cannot be attributed to a simple statistical fluctuation. In the previous example, the value $x_s = 1.5$ cannot be eliminated, according to the Chauvenet criterion.

Problems

9.1. The oscillation period of an elastic spring has been measured $N = 2$ times, and the following values have been obtained (in seconds),
    0.363, 0.362 .
Estimate the mean of the parent distribution $\tilde{m}$ and the standard deviation of the distribution of sample means $\tilde{\sigma}[m^*]$. Calculate the confidence level for the coverage factors $k = 1$ and $k = 3$.

Eight other measurements are performed, and the total $N = 10$ resulting values are:
    0.363, 0.362, 0.364, 0.365, 0.365, 0.367, 0.360, 0.363, 0.361, 0.364 .
Again, estimate $\tilde{m}$ and $\tilde{\sigma}[m^*]$, and calculate the confidence level for the coverage factors $k = 1$ and $k = 3$. Compare with the case $N = 2$.

9.2. The specific heat of iron has been measured $N = 4$ times by a group of students, and the following values have been obtained (in $J\,kg^{-1}\,K^{-1}$),
    450, 449, 431, 444 .
Estimate the mean of the parent distribution $\tilde{m}$ and the standard deviation of the distribution of sample means $\tilde{\sigma}[m^*]$. The value quoted in a physics handbook, $450\ J\,kg^{-1}\,K^{-1}$, is assumed as the true value m. Evaluate if the relative discrepancy $(m^* - m)/\tilde{\sigma}[m^*]$ can reasonably be attributed to random fluctuations.

10 Correlation of Physical Quantities

Accurate measurements of single quantities are relevant in both science and technology. Continuous efforts are, for example, devoted to reduce the uncertainty of the fundamental constants of physics (Appendix C.2).

Also of basic importance is the measurement of two or more correlated quantities and the search for functional relations between their values that can lead to the formulation of physical laws. For example, the measurement of the intensity of the electrical current I flowing in a metallic conductor as a function of the difference of potential V leads to a proportionality relation that is known as Ohm law.

The uncertainty that affects the value of any measured physical quantity plays an important role when one tries to establish correlations between two or more quantities, and express them by functional relations. The uncertainties of single measures necessarily lead to uncertainties of the functional relations between quantities. In this chapter, some techniques are introduced, useful to recognize the existence of a correlation between two quantities, and to express it in a functional form, properly taking into account the measurement uncertainties.

10.1 Relations Between Physical Quantities

Let us consider two physical quantities X and Y. $\mathcal{N}$ different values x_i of X are measured, and for each value the uncertainty δx_i is evaluated, for example, by repeating the measurement many times. In correspondence with each value x_i of X, a value y_i of Y is measured, and the uncertainty δy_i is evaluated. The procedure leads to $\mathcal{N}$ pairs of values

$$x_i \pm \delta x_i, \qquad y_i \pm \delta y_i, \qquad (i = 1, \ldots, \mathcal{N}). \qquad (10.1)$$

The pairs of values can be listed in a table. A more effective way for identifying and studying the possible correlation between the values of X and Y is represented by a graph, where the pairs of central values (x_i, y_i) are represented by points, and the uncertainties δx_i and δy_i are represented by error bars, horizontal and vertical, respectively. Useful details on tables and graphs are given in Appendices A.2 and A.3, respectively.

178   10 Correlation of Physical Quantities

The following examples, illustrated by the graphs of Fig. 10.1, clarify some typical situations.

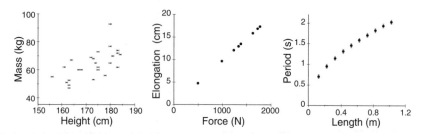

Fig. 10.1. Correlation between physical quantities. Left: the mass of a group of persons as a function of their height (Example 10.1). Center: elongation of an elastic spring as a function of the applied force (Example 10.2). Right: period of a pendulum as a function of its length (Example 10.3). The error bars, when not visible, are smaller than the symbol size.

Example 10.1. $\mathcal{N}$ persons are subjected to a medical examination. Height $h_i \pm \delta h_i$ and mass $m_i \pm \delta m_i$ of each person are measured ($i = 1, \ldots, \mathcal{N}$). In the graph of Fig. 10.1 (left), the points representing the pairs (h_i, m_i) are very dispersed. One can, however, notice a weak correlation between mass and height: when the height increases, in general the mass increases as well. A probabilistic procedure for evaluating the degree of correlation between two quantities is introduced in Sect. 10.2.

Example 10.2. An elastic spring is suspended from a fixed point. A variable force is applied to the free end. To each value $F_i \pm \delta F_i$ of the force corresponds a value $y_i \pm \delta y_i$ of deformation of the spring. From the graph of Fig. 10.1 (center), not only is it immediate to recognize a correlation between the two quantities F and y, but it appears also reasonable to hypothesize that the two quantities are connected by a proportionality relation $y = BF$, where B is a constant. A general procedure to evaluate the coefficients A and B of a linear relation $Y = A + BX$ is introduced in Sect. 10.3.

Example 10.3. The distance ℓ of a simple pendulum from its suspension point is varied. For each value of length $\ell_i \pm \delta \ell_i$, a different value of period $T_i \pm \delta T_i$ is measured. Also in this case, from the graph of Fig. 10.1 (right), it is reasonable to hypothesize that the two quantities ℓ and T are connected by a functional relation that, however, is clearly not linear. Some procedures for treating nonlinear relations between physical quantities are considered in the following sections.

10.2 Linear Correlation Coefficient

In this section, a method is introduced for quantitatively evaluating the degree of correlation between two physical quantities X and Y. Reference can be made to Example 10.1 and to Fig. 10.1, left.

Sample Variance and Covariance

Once the N pairs of values (x_j, y_j) have been measured, the sample means, the sample variances, and the sample standard deviations can be calculated as

$$m_x^* = \frac{1}{N}\sum_j x_j, \quad D_x^* = \frac{1}{N}\sum_j (x_j - m_x^*)^2, \quad \sigma_x^* = \sqrt{D_x^*}, \quad (10.2)$$

$$m_y^* = \frac{1}{N}\sum_j y_j, \quad D_y^* = \frac{1}{N}\sum_j (y_j - m_y^*)^2, \quad \sigma_y^* = \sqrt{D_y^*}. \quad (10.3)$$

The sample covariance is

$$\sigma_{xy}^* = \frac{1}{N}\sum_j (x_j - m_x^*)(y_j - m_y^*). \quad (10.4)$$

It is important to note the different meaning of variance and covariance in the previous application of Sect. 8.4 and in the present application. In Sect. 8.4, variance and covariance refer to the random fluctuations of two given values x_i and y_i with respect to their means, and are connected to uncertainty. Here, variance and covariance refer to N pairs (x_i, y_i) of different values of the two quantities.

Definition of the Linear Correlation Coefficient

The linear correlation coefficient r of the two quantities X and Y is defined as follows:

$$r = \frac{\sigma_{xy}^*}{\sigma_x^* \sigma_y^*} = \frac{\sum_j (x_j - m_x^*)(y_j - m_y^*)}{\sqrt{\sum_j (x_j - m_x^*)^2}\sqrt{\sum_j (y_j - m_y^*)^2}}. \quad (10.5)$$

As has already been noticed in Sect. 8.4, the Schwartz inequality ensures that $|\sigma_{xy}^*| \le \sigma_x^* \sigma_y^*$. This means that the linear correlation coefficient cannot be smaller than -1 nor larger than $+1$:

$$|r| \le 1. \quad (10.6)$$

The Case of Perfect Linear Correlation

To understand the meaning of the linear correlation coefficient r, it is convenient to first consider a particular case. Let us suppose that the quantities X and Y are exactly connected by a linear relation:

$$Y = A + BX. \tag{10.7}$$

In this case, for any pair of values (x_i, y_i) and for the sample means m_x^*, m_y^*, the following relations are true,

$$y_i = A + Bx_i, \tag{10.8}$$
$$m_y^* = A + Bm_x^*. \tag{10.9}$$

By subtracting (10.9) from (10.8), one gets

$$y_i - m_y^* = B(x_i - m_x^*). \tag{10.10}$$

Taking into account (10.10), one can easily verify that, for an exact linear relation, the expression of the linear correlation coefficient (10.5) is simply:

$$r = \frac{B}{\sqrt{B}} = \frac{B}{|B|} = \begin{cases} +1 & \text{if } B > 0, \\ -1 & \text{if } B < 0. \end{cases} \tag{10.11}$$

The two extremum values $r = 1$ and $r = -1$ thus correspond to a perfectly linear relation, with angular coefficient B positive and negative, respectively.

Interpretation of the Linear Correlation Coefficient

Let us now try to understand how a generic value $-1 < r < 1$ of the linear correlation coefficient, obtained through (10.5) from N pairs of values (x_j, y_j), can be interpreted. It is worth noting that the coefficient r is calculated from an experimental sample of finite size, subject to random fluctuations. The correlation coefficient r is thus a random variable.

Let us hypothesize at first that the two quantities X and Y are totally uncorrelated. One expects then that, when the number N of measured pairs increases, the covariance σ_{xy}^* tends to stabilize around zero: for $N \to \infty$, $\sigma_{xy}^* \to 0$. Actually, for a finite number N of pairs, the sample covariance σ_{xy}^* is generally not zero, also when the two quantities are completely uncorrelated, its value depending on the sample. Let r_o be the linear correlation coefficient calculated through (10.5) (the index "o" here stands for "observed"). The quantity r_o is a random variable. The distribution of the random variable r_0 can be calculated as a function of the number N of measured pairs (x_j, y_j). In Appendix C.6, one finds a table containing the integrals of the distribution, for selected values of N and r_o. The integrals correspond to the probability

$$\mathcal{P}_{\mathcal{N}}\left(|r| \geq r_{o}\right) \tag{10.12}$$

that the correlation coefficient r is larger than the observed value r_o, for a pair of completely uncorrelated quantities X and Y.

Let us now consider a realistic case. $\mathcal{N}$ pairs of values (x_j, y_j) of two quantities X and Y have been measured, and one wants to evaluate their degree of correlation. To this purpose, one calculates the observed correlation coefficient r_o through (10.5). One then looks in a table of probability values for the distribution of r (Appendix C.6).

- If $\mathcal{P}_{\mathcal{N}}(|r| \geq r_o) < 5\%$, the correlation between the two quantities is said to be *significant*.
- If $\mathcal{P}_{\mathcal{N}}(|r| \geq r_o) < 1\%$, the correlation between the two quantities is said to be *very significant*.

10.3 Linear Relations Between Two Quantities

Let us now consider the case of Example 10.2 (Fig. 10.1, center), where it is very reasonable to assume that a proportionality relation exists between the two measured quantities. For the sake of completeness, we consider the more general case of a linear relation:

$$Y = A + BX . \tag{10.13}$$

Two problems have to be solved:

(a) To determine the parameters A and B of the straight line $Y = A + BX$ best fitting the experimental points. This problem is solved in this section.
(b) To evaluate the degree of reliability of the hypothesis that a linear relation actually fits the experimental points. This problem, the test of the hypothesis, is solved in Chap. 11.

A first evaluation of the parameters A and B and of the order of magnitude of their uncertainties δA and δB can often be obtained by the following graphical method. One draws the straight lines of maximum and minimum slope consistent with the uncertainty crosses of the experimental points (Fig. 10.2). The equations of the two straight lines are

$$Y = A_1 + B_1 X , \qquad Y = A_2 + B_2 X . \tag{10.14}$$

It is reasonable to assume the values of the parameters A and B as

$$A = \frac{A_1 + A_2}{2} , \qquad B = \frac{B_1 + B_2}{2} , \tag{10.15}$$

and to evaluate the corresponding uncertainties through the approximate relations

$$\delta A \simeq \frac{|A_1 - A_2|}{2} , \qquad \delta B \simeq \frac{|B_1 - B_2|}{2} . \tag{10.16}$$

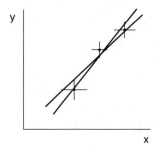

Fig. 10.2. Straight lines of maximum and minimum slope compatible with the uncertainty crosses.

Linear Regression

An effective procedure for analytically calculating the coefficients A and B of the straight line $A + BX$ is represented by *linear regression*, based on the *least squares method*. The procedure consists of determining the values of the parameters A and B that minimize the global discrepancy between the experimental points (x_i, y_i) and the straight line $Y = A + BX$, taking into account the uncertainty of each experimental point.

For each point, the discrepancy is measured along the y-axis, considering the square of the difference between the vertical coordinate of the point and the straight line (Fig. 10.3, left):

$$(y_i - A - Bx_i)^2 \ . \tag{10.17}$$

The least squares method takes into account only the uncertainties δy_i on the Y quantities. We suppose, for the moment, that the uncertainties δx_i are actually negligible; below, we learn how to take into account the uncertainties δx_i as well.

For each point, the discrepancy (10.17) is divided by the uncertainty $(\delta y_i)^2$, say multiplied by the weight $w_i = 1/(\delta y_i)^2$ (Fig. 10.3, right). The global discrepancy is measured by the sum over all points:

$$\chi^2 = \sum_{i=1}^{N} \frac{(y_i - A - Bx_i)^2}{(\delta y_i)^2} = \sum_{i=1}^{N} w_i \, (y_i - A - Bx_i)^2 \ . \tag{10.18}$$

By convention, the sum is named χ^2 (chi square). The values x_i, y_i and δy_i are known, therefore the quantity χ^2 can be considered a function of the two variables A and B. The goal is thus to determine analytically the values of A and B that minimize the value of χ^2.

Let us first consider the case of *direct proportionality*

$$Y = BX \ . \tag{10.19}$$

The global discrepancy is measured by the sum

10.3 Linear Relations Between Two Quantities

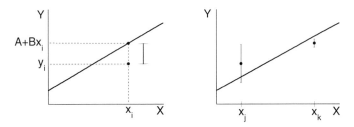

Fig. 10.3. Left: Discrepancy between an experimental point and the theoretical straight line, in correspondence of a given value x_i. Right: The absolute discrepancy is larger for the point at x_j than for the point at x_k; the discrepancy divided by the uncertainty is instead smaller for the point at x_j than for the point at x_k.

$$\chi^2 = \sum_{i=1}^{N} \frac{(y_i - Bx_i)^2}{(\delta y_i)^2} = \sum_{i=1}^{N} w_i \, (y_i - Bx_i)^2 \; . \tag{10.20}$$

The quantity χ^2 is a function of the variable B. A necessary condition for χ^2 to be minimum is that the first derivative with respect to B be zero:

$$\frac{\mathrm{d}\chi^2}{\mathrm{d}B} = 0 \; . \tag{10.21}$$

One can easily verify that (10.21) is satisfied by

$$B = \frac{\sum_i w_i x_i y_i}{\sum_i w_i x_i^2} \; , \tag{10.22}$$

and the value B of (10.22) actually corresponds to a minimum of χ^2.

If the uncertainties δy_i are the same for all points, (10.22) becomes

$$B = \frac{\sum_i x_i y_i}{\sum_i x_i^2} \; . \tag{10.23}$$

Let us now consider the *general case of linearity* (Fig. 10.4).

$$Y = A + BX \; . \tag{10.24}$$

The quantity χ^2 (10.18) is now a function of two variables, A and B. A necessary condition for χ^2 to be minimum is that its first derivatives with respect to A and B be zero:

$$\begin{cases} \dfrac{\partial \chi^2}{\partial A} = 0 \\ \dfrac{\partial \chi^2}{\partial B} = 0 \end{cases} \Rightarrow \begin{cases} (\sum_i w_i) \, A + (\sum_i w_i x_i) \, B = \sum_i w_i y_i \; , \\ (\sum_i w_i x_i) \, A + (\sum_i w_i x_i^2) \, B = \sum_i w_i x_i y_i \; . \end{cases} \tag{10.25}$$

One thus obtains a system of two linear equations with unknowns A and B, whose solution is

$$A = \frac{\left(\sum_i w_i x_i^2\right)\left(\sum_i w_i y_i\right) - \left(\sum_i w_i x_i\right)\left(\sum_i w_i x_i y_i\right)}{\Delta_w}, \quad (10.26)$$

$$B = \frac{\left(\sum_i w_i\right)\left(\sum_i w_i x_i y_i\right) - \left(\sum_i w_i y_i\right)\left(\sum_i w_i x_i\right)}{\Delta_w}, \quad (10.27)$$

where

$$\Delta_w = \left(\sum_i w_i\right)\left(\sum_i w_i x_i^2\right) - \left(\sum_i w_i x_i\right)^2. \quad (10.28)$$

One can easily verify that the values A and B of (10.26) and (10.27) actually correspond to a minimum of χ^2.

If the uncertainties δy_i are the same for all points, (10.26) and (10.27) become

$$A = \frac{\left(\sum_i x_i^2\right)\left(\sum_i y_i\right) - \left(\sum_i x_i\right)\left(\sum_i x_i y_i\right)}{\Delta}, \quad (10.29)$$

$$B = \frac{N\left(\sum_i x_i y_i\right) - \left(\sum_i y_i\right)\left(\sum_i x_i\right)}{\Delta}, \quad (10.30)$$

where

$$\Delta = N\left(\sum_i x_i^2\right) - \left(\sum_i x_i\right)^2. \quad (10.31)$$

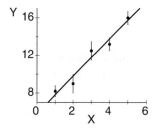

Fig. 10.4. Example of linear regression: the continuous line is the straight line best fitting the experimental points.

Taking into Account the Uncertainties of X

If the uncertainties δx_i are not negligible, they can be taken into account by a procedure based on the following steps.

1. Two approximate values A' and B' of the straight line parameters are evaluated by the graphical method (10.15) or through (10.29)–(10.31).
2. The uncertainties $(\delta x_i)_{\text{exp}}$ of X are transformed into contributions $(\delta y_i)_{\text{tra}}$ to the uncertainties of Y by the propagation procedures of Chap. 8, obtaining, for each point

$$(\delta y_i)_{\text{tra}} = |B'|(\delta x_i)_{\text{exp}}. \quad (10.32)$$

3. The two contributions to the uncertainty of Y, the experimental one and the transferred one, are quadratically summed for each point:

$$(\delta y_i)^2_{\text{tot}} = (\delta y_i)^2_{\text{exp}} + (\delta y_i)^2_{\text{tra}} . \tag{10.33}$$

4. The final parameters A and B are calculated through (10.26)–(10.28), where now $w_i = 1/(\delta y_i)^2_{\text{tot}}$.

Uncertainties of the Parameters A and B

Because the values y_i are affected by uncertainties δy_i, one expects that the parameters A and B are affected by uncertainty as well. To evaluate the uncertainties δA and δB, the propagation procedures of Chap. 8 are used.

Let us first consider the simpler case of direct proportionality, $Y = BX$. Equation (10.22) shows that B is a linear combination of y_i values, as in (8.1):

$$B = \sum_i \beta_i y_i , \qquad \text{where } \beta_i = \frac{w_i x_i}{\sum_i w_i x_i^2} . \tag{10.34}$$

One can then apply the propagation rule (10.18), with $(\delta y_i)^2 = 1/w_i$:

$$(\delta B)^2 = \sum_i \beta_i^2 (\delta y_i)^2 = \frac{\sum_i w_i^2 x_i^2 / w_i}{(\sum_i w_i x_i^2)^2} = \frac{1}{\sum_i w_i x_i^2} . \tag{10.35}$$

If all uncertainties δy_i are equal, (10.35) reduces to

$$(\delta B)^2 = \frac{\sum_i x_i^2}{(\delta y)^2} . \tag{10.36}$$

Let us now consider the general case of linearity. Both parameters A and B, (10.26) and (10.27), are linear combinations of the values y_i, as in (8.1):

$$A = \sum_i \alpha_i y_i, \quad \text{where } \alpha_i = \frac{w_i(\sum_i w_i x_i^2) - w_i x_i(\sum_i w_i x_i)}{\Delta_w} , \tag{10.37}$$

$$B = \sum_i \beta_i y_i, \quad \text{where } \beta_i = \frac{w_i x_i(\sum_i w_i) - w_i(\sum_i w_i x_i)}{\Delta_w} . \tag{10.38}$$

One can again use the propagation rule (8.6), with $(\delta y_i)^2 = 1/w_i$, and taking into account that Δ_w is given by (10.28), one can verify that

$$(\delta A)^2 = \sum_i \alpha_i^2 (\delta y_i)^2 = \frac{\sum_i w_i x_i^2}{\Delta_w} , \tag{10.39}$$

$$(\delta B)^2 = \sum_i \beta_i^2 (\delta y_i)^2 = \frac{\sum_i w_i}{\Delta_w} . \tag{10.40}$$

If all the uncertainties δy_i are equal, (10.39) and (10.40) reduce to

$$(\delta A)^2 = \frac{\sum_i x_i^2}{\Delta} (\delta y)^2 , \qquad (\delta B)^2 = \frac{N}{\Delta} (\delta y)^2 , \tag{10.41}$$

where Δ is given by (10.31).

Linearization of Nonlinear Relations

It is quite easy to check, by visual inspection of a graph, if $\mathcal{N}$ experimental points (x_i, y_i) are approximately arranged along a straight line (Fig. 10.1, center); in this case, the parameters of the straight line can be determined by the linear regression procedure. It can instead be difficult to recognize, by visual inspection, nonlinear functional relations (Fig. 10.1, right). It is, for example, difficult to distinguish between a parabolic and a cubic behavior, $Y = \alpha X^2$ and $Y = \beta Y^3$, respectively.

In some cases, simple transformations of the graph axes allow the transformation of a generic function $Y = \phi(X)$ to a straight line $Y = A + BX$, to which the linear regression procedure can be applied. For example, a parabolic behavior $Y = \alpha X^2$ can be transformed into a linear behavior by plotting the Y values as a function of $Z = X^2$. Other very effective linearization procedures are based on the use of logarithmic graphs, and are summarized in Appendix A.3. Working examples can be found in Experiments E.4, E.5, and E.7 of Appendix E.

10.4 The Least Squares Method

In Sect. 10.3, the linear regression procedure was founded on the least squares method. In this section, a justification of the least squares method, based on the criterion of maximum likelihood (Sect. 7.3), is given, and its application extended from the case of a linear function $Y = A + BX$ to the case of a generic function $Y = \phi(X)$.

Introduction to the Problem

Let us suppose that $\mathcal{N}$ pairs of values $(x_i \pm \delta x_i, y_i \pm \delta y_i)$ of two physical quantities X and Y have been measured. A functional relation $Y = \phi(X)$, that satisfactorily describes the relation between X and Y, is sought. In general, the function $Y = \phi(X)$ depends on several parameters $\lambda_1, \lambda_2, \ldots$ so that it will be expressed as $Y = \phi(X, \{\lambda_k\})$. For example, in the case of a linear dependence $Y = A + BX$ (Sect. 10.3), there are two parameters $\lambda_1 = A, \lambda_2 = B$.

The procedure for fitting a function $Y = \phi(X, \{\lambda_k\})$ to a set of $\mathcal{N}$ points $(x_i \pm \delta x_i, y_i \pm \delta y_i)$ can be decomposed into two steps:

1. The choice of the form of the function $Y = \phi(X, \{\lambda_k\})$: linear, parabolic, exponential, sinusoidal, and so on
2. The evaluation of the parameters $\{\lambda_k\}$

As for the first step, let us suppose that the form of the function $Y = \phi(X, \{\lambda_k\})$ is known from independent considerations of theoretical or experimental nature. The goodness of the hypothesis $Y = \phi(X, \{\lambda_k\})$ can be

10.4 The Least Squares Method

a posteriori evaluated by means of tests, such as the chi square test considered in Chap. 11.

The second step corresponds to the problem of the estimation of the parameters of a parent population from a finite sample (Sect. 7.3), and the method of least squares can be derived from the maximum likelihood criterion.

Maximum Likelihood and Least Squares

The derivation of the least squares method from the maximum likelihood criterion is based on two hypotheses.

1. The uncertainty of X is negligible with respect to the uncertainty of Y, so that one can assume $\delta x_i = 0$, $\delta y_i \neq 0$ for each point. This hypothesis is not particularly restrictive; the possibility of taking into account nonzero values of δx_i has already been considered in the previous section for the linear case, and is generalized below.
2. The uncertainty of Y is expressed by the standard deviation of a suitable normal distribution, $\delta y_i = \sigma_i$. In the following, to simplify the notation, the symbol σ_i is used throughout. For example, for random fluctuations, σ_i here stands for $\tilde{\sigma}_i[m_i^*]$.

To introduce the maximum likelihood criterion, the same approach is used as in previous applications of Sect. 7.3.

If the parameters $\{\lambda_k\}$ of $\phi(X, \{\lambda_k\})$ were known, once a value x_i has been given, the probability density of the corresponding value y_i would be

$$f(y_i) = \frac{1}{\sigma_i \sqrt{2\pi}} \exp\left\{-\frac{[y_i - \phi(x_i, \{\lambda_k\})]^2}{2\sigma_i^2}\right\}. \quad (10.42)$$

Once a set of $\mathcal{N}$ values $x_1, x_2, \ldots, x_\mathcal{N}$ has been given, the probability density of obtaining $\mathcal{N}$ independent values $y_1, y_2, \ldots, y_\mathcal{N}$ would be a multivariate density that can be factorized into the product of univariate densities $f(y_i)$:

$$g(y_1, y_2, \ldots, y_\mathcal{N}; \{\lambda_k\}) = \prod_{i=1}^{\mathcal{N}} \frac{1}{\sigma_i \sqrt{2\pi}} \exp\left\{-\frac{[y_i - \phi(x_i, \{\lambda_k\})]^2}{2\sigma_i^2}\right\}$$

$$= \frac{1}{(\prod_i \sigma_i)(2\pi)^{\mathcal{N}/2}} \exp\left\{-\frac{1}{2}\sum_{i=1}^{\mathcal{N}} \frac{[y_i - \phi(x_i, \{\lambda_k\})]^2}{\sigma_i^2}\right\}. \quad (10.43)$$

In the real cases, the $\mathcal{N}$ values $y_1, y_2, \ldots, y_\mathcal{N}$ are known, whereas the values of the parameters $\{\lambda_k\}$ are unknown. According to the maximum likelihood criterion, the best estimates of the parameters $\{\lambda_k\}$ are the values that maximize the probability density $g(y_1, y_2, \ldots, y_\mathcal{N}; \{\lambda_k\})$ in (10.43).

Finding the maximum of $g(y_1, y_2, \ldots, y_\mathcal{N}; \{\lambda_k\})$ with respect to the parameters $\{\lambda_k\}$ corresponds to finding the minimum of the sum in the exponent in the last member of (10.43). The sum is indicated by convention as χ^2 (chi square):

$$\chi^2 = \sum_{i=1}^{N} \frac{[y_i - \phi(x_i, \{\lambda_k\})]^2}{\sigma_i^2} \ . \tag{10.44}$$

Each term in the sum (10.44) measures the discrepancy between experimental points and theoretical curve, compared with the uncertainty σ_i. Because the values x_i, y_i, and σ_i are known, χ^2 is a function only of the p parameters $\{\lambda_k\}$,

$$\chi^2 = \chi^2(\lambda_1, \lambda_2, \ldots, \lambda_p) \ , \tag{10.45}$$

defined in a p-dimensional space.

To determine the parameters $\{\lambda_k\}$ best fitting the experimental points, it is necessary to find the absolute minimum of the function χ^2 in the p-dimensional parameter space. The minimization procedure can be performed in various ways, depending on the form of the function $\phi(X, \{\lambda_k\})$.

Minimization for Linear Functions

The minimization procedure of χ^2 can be analytically performed if the function $\phi(X, \{\lambda_k\})$ linearly depends on the parameters $\{\lambda_k\}$. The most general linear expression of ϕ is

$$\begin{aligned}\phi(X, \{\lambda_k\}) &= \lambda_1\, h_1(X) + \lambda_2\, h_2(X) + \lambda_3\, h_3(X) + \cdots \\ &= \sum_{k=1}^{p} \lambda_k h_k(X) \ ,\end{aligned} \tag{10.46}$$

where the h_k are known functions of the variable X. The χ^2 is then

$$\chi^2 = \sum_{i=1}^{N} \frac{1}{\sigma_i^2} \left[y_i - \lambda_1 h_1(x_i) - \lambda_2 h_2(x_i) - \lambda_3 h_3(x_i) - \cdots \right]^2 \ . \tag{10.47}$$

The minimum of χ^2 is obtained by imposing that all its first derivatives with respect to the p parameters λ_k are zero. One can easily verify that, for any parameter λ_k, one obtains a first-degree equation with p unknowns λ_k:

$$\frac{\partial \chi^2}{\partial \lambda_k} = \sum_{i=1}^{N} \frac{2}{\sigma_i^2}\, h_k(x_i)\, [y_i - \lambda_1 h_1(x_i) - \lambda_2 h_2(x_i) - \lambda_3 h_3(x_i) - \cdots] = 0 \ . \tag{10.48}$$

By considering the derivatives with respect to all the parameters λ_k, one obtains a system of p linear equations with constant coefficients and p unknowns λ_k. To determine the parameters λ_k, one can use any of the well-established techniques for the solution of systems of equations, such as the determinant method.

The linear function $Y = A + BX$ of Sect. 10.3 is a simple case of (10.46), with $h_1(X) = 1$, $h_2(X) = X$, $h_3(X) = h_4(X) = \cdots = 0$. The linear regression procedure of Sect. 10.3 is the simplest application of the least squares method.

The linear regression case can be generalized to a generic polynomial function

$$Y = \lambda_1 + \lambda_2 X + \lambda_3 X^2 + \lambda_4 X^3 + \cdots \qquad (10.49)$$

that is again a particular case of (10.46). A polynomial of degree $p-1$ has p parameters. By imposing that the first derivatives of χ^2 with respect to the parameters are zero,

$$\frac{\partial \chi^2}{\partial \lambda_k} = \frac{\partial}{\partial \lambda_k} \sum_{i=1}^{N} \frac{1}{\sigma_i^2} [y_i - \lambda_1 - \lambda_2 x_i - \lambda_3 x_i^2 - \lambda_4 x_i^3 - \cdots]^2 = 0, \qquad (10.50)$$

one obtains a system of p linear equations with constant coefficients and p unknowns. The solution of the system gives the values of the p parameters $\lambda_1, \lambda_2, \ldots, \lambda_p$. The procedure is called *polynomial regression* (Fig. 10.5).

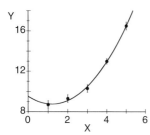

Fig. 10.5. Example of polynomial regression: the continuous line is the second-degree polynomial best fitting the experimental points.

The parameters $\{\lambda_k\}$, analytically obtained by the regression procedures, are a function of the values x_i, y_i and of the uncertainties σ_i. One can thus evaluate their uncertainties $\{\delta \lambda_k\}$ by the propagation procedures of Chap. 8, as was done in Sect. 10.3 for the particular case of linear regression.

Note. It is important to distinguish the linearity of the function $\phi(X, \{\lambda_k\})$ with respect to the parameters $\{\lambda_k\}$, which allows an analytical solution of the minimization procedure, from the linearity with respect to the variable X, which leads to the particular case of linear regression.

Minimization for nonlinear functions

Let us now consider the functions $\phi(X, \{\lambda_k\})$ that are nonlinear with respect to the parameters $\{\lambda_k\}$.

In some cases, one can obtain a linearization of the function by suitable changes of variables, and then again apply the analytical methods of regression.

Example 10.4. An exponential relation $Y = a \exp(bX)$ can be linearized by taking the natural logarithm: $\ln Y = \ln a + bX$. The new expression is linear with respect to the parameters $\lambda_1 = \ln a$ and $\lambda_2 = b$. By substituting $Y' = \ln Y$ one obtains the expression $Y' = \lambda_1 + \lambda_2 X$, which is also linear with respect to X, and allows the application of the linear regression procedure.

Example 10.5. A generic power law $Y = aX^b$ can be linearized by taking the natural or decimal logarithms: $\log Y = \log a + b \log X$. The new expression is linear with respect to the parameters $\lambda_1 = \log a$ and $\lambda_2 = b$. By substituting $Y' = \log Y$ and $X' = \log X$, one obtains the expression $Y' = \lambda_1 + \lambda_2 X'$, that is also linear with respect to X', and allows the application of the linear regression procedure.

If the function $\phi(X, \{\lambda_k\})$ cannot be linearized with respect to the parameters $\{\lambda_k\}$, the minimization of χ^2 cannot be performed by analytical procedures, and requires numerical approaches.

Example 10.6. An example of a function that cannot be linearized is the law of the damped harmonic oscillator, $Y = A \exp(-\gamma X) \sin(\omega_0 X + \vartheta)$.

The simplest method of numerical minimization consists of creating a regular lattice in the p-dimensional space of the parameters $\{\lambda_k\}$. Each lattice point corresponds to a set of p values $\{\lambda_1, \lambda_2, \ldots, \lambda_p\}$. For each lattice point, one calculates the χ^2 value through (10.44), and then one searches the point corresponding to the minimum of χ^2. The search can be made progressively more precise by increasing the density of the lattice points in selected regions of the parameter space.

This method, in principle particularly simple, can become very slow and ineffective when the number p of parameters $\{\lambda_k\}$ increases: if n values are considered for each parameter, the total number of lattice points to be explored is n^p, a number that can easily become exceedingly large. Various alternative approaches, faster and more effective, have been developed for searching the absolute minimum of χ^2, and are implemented in easily available computer codes.

Taking into Account the Uncertainties of X.

The least squares method is based on the hypothesis that the uncertainty of the independent variable X is negligible. If the uncertainties δx_i are not negligible, one can take them into account by the following procedure.

1. A preliminary application of the least squares method is made, taking into account only the uncertainties $\sigma_i = \delta y_i$. One obtains approximate values $\{\lambda'_k\}$ of the parameters, and a first approximate evaluation of the function $\phi'(X, \{\lambda'_k\})$. (For linear regression, this first step can be made graphically, as shown in Sect. 10.3.)

2. The uncertainties $(\delta x_i)_{\exp}$ are transformed into contributions $(\delta y_i)_{\text{tra}}$ to the uncertainties of Y, by the propagation procedures of Chap. 8, and one obtains, for each point

$$(\delta y_i)_{\text{tra}} = \left|\frac{d\phi'(X, \{\lambda'_k\})}{dX}\right|_{x_i} (\delta x_i)_{\exp} . \tag{10.51}$$

3. The two contributions to the uncertainty of Y, the experimental one and the transferred one, are quadratically summed for each point:

$$(\delta y_i)^2_{\text{tot}} = (\delta y_i)^2_{\exp} + (\delta y_i)^2_{\text{tra}} . \tag{10.52}$$

4. The definitive application of the least squares method is made, where now the uncertainties are $\sigma_i = (\delta y_i)_{\text{tot}}$.

A Posteriori Evaluation of the Uncertainty on Y

In some cases, it happens that the uncertainties δy_i are underestimated. Typical cases are:

(a) Each value y_i is the outcome of only one measurement, and the quoted uncertainty δy_i only takes into account resolution, not the possible effects of random fluctuations.
(b) The uncertainty due to uncompensated systematic errors has not been taken into account.

In such cases, if there are good reasons to trust the hypothesis of the functional relation $Y = \phi(X, \{\lambda_k\})$, for example from theoretical considerations or from the results of independent experiments, then the parameters $\{\lambda_k\}$ can still be estimated, and one can attempt an evaluation a posteriori of the average uncertainty $\delta y = \sigma$.

The parameters $\{\lambda_k\}$ are calculated by minimizing the sum

$$\psi^2 = \sum_{i=1}^{N} [y_i - \phi(X, \{\lambda_k\})]^2 , \tag{10.53}$$

that corresponds to minimizing the sum (10.44) if the uncertainties σ_i are are equal.

The unknown average uncertainty σ can be now evaluated, again using the maximum likelihood criterion, by maximizing the probability density

$$G_{\mathcal{N}}(y_1, y_2, \ldots, y_N; \sigma) = \prod_{i=1}^{N} \frac{1}{\sigma\sqrt{2\pi}} \exp\left\{-\frac{[y_i - \phi(X, \{\lambda_k\})]^2}{2\sigma^2}\right\} \tag{10.54}$$

with respect to σ^2. One can easily verify that the minimum of the exponent in (10.54) is obtained for

$$\sigma^2 = \frac{1}{N} \sum_i [y_i - \phi(X, \{\lambda_k\})]^2 . \tag{10.55}$$

The estimate of uncertainty given by (10.55) is not unbiased (Sect. 7.3). Let us, for example, consider the limiting case of N experimental points to which a polynomial with $p = N$ parameters is fitted. In such a case, the curve determined by the polynomial regression procedure would exactly contain all the experimental points, and one would obtain $\sigma^2 = 0$ from (10.55).

Because the p parameters are determined from the experimental data, only $N - p$ independent values are available to evaluate the uncertainty. The problem is said to have $N - p$ degrees of freedom. As a consequence, an unbiased estimate of the uncertainty can be obtained only if $N > p$, and is given by

$$\sigma^2 = \frac{1}{N - p} \sum_i [y_i - \phi(X, \{\lambda_k\})]^2 . \tag{10.56}$$

Problems

Problem 1. The elongation x of a spring has been measured as a function of the applied force F (Experiment E.3 of Appendix E). The following table lists the measured pairs of values (F_j, x_j). The uncertainties are $\delta F = 3 \times 10^{-4}$ N and $\delta x = 0.4$ mm.

F_j (N)	0.254	0.493	0.739	0.986	1.231	1.477	1.721
x_j (mm)	25	49	73	97	121	146	170

Plot the experimental points on a linear graph. Evaluate the best fitting straight line $F = Bx$ by the linear regression procedure, and plot it on the same graph of the experimental points. Calculate the uncertainty of the parameter B.

Problem 2. The period T of a pendulum has been measured as a function of the length ℓ (Experiment E.5 of Appendix E). The following table lists the measured pairs of values (ℓ_j, T_j). The uncertainties are $\delta\ell = 6 \times 10^{-4}$ m and $\delta T = 1.5 \times 10^{-4}$ s.

ℓ_j (m)	1.194	1.063	0.962	0.848	0.737	0.616	0.504	0.387	0.270	0.143
T_j (s)	2.198	2.072	1.974	1.856	1.733	1.580	1.429	1.255	1.048	0.766

Plot the experimental points on a linear graph, and verify that the relation between T and ℓ is not linear.

Plot the values $Y = T^2$ against ℓ, and evaluate the best fitting straight line $Y = B\ell$ by the linear regression procedure, and plot it on the same graph of the experimental points. Calculate the uncertainty of the parameter B.

11 The Chi Square Test

A central problem of statistical data analysis is represented by the test of theoretical hypotheses against experimental data. Typical examples are the test of consistency of a normal distribution with a histogram of measured values (Sect. 4.3), or the test of consistency of a functional relation with a set of experimental points (Sect. 10.3).

Several different procedures have been devised for performing such types of tests. One of the most frequently used is the chi square test (after the name of the Greek letter χ, "chi") to which this chapter is dedicated.

11.1 Meaning of the Chi Square Test

The chi square test gives a criterion for verifying, on probabilistic grounds, the consistency of a theoretical hypothesis with a set of experimental data. In this chapter, the chi square test is studied for two types of problems:

(a) Comparison between a finite set of values of a random variable and a probability distribution
(b) Comparison between a set of pairs of measured values (x_k, y_k) and a functional relation $Y = \phi(X)$

Comparison Between Expected and Observed Values

As a first step, let us introduce a general procedure for measuring the discrepancy between observed experimental data and theoretical expected values. The topic can be better introduced by some simple examples.

Example 11.1. Binomial distribution. Twenty dice ($n = 20$) are contemporarily tossed N times. At each toss, the number of dice exhibiting the face "2" is a discrete random variable K, whose possible values are $0 \leq k \leq 20$. For N tosses, a number O_k of observed outcomes is associated with each value of K. If the dice are perfectly symmetrical, the random variable K is distributed according to the binomial law $\mathcal{P}_{np}(k) = \mathcal{P}_{20,1/6}(k)$, and the expected number of outcomes is $E_k = N\mathcal{P}_{20,1/6}(k)$ for each value of K. The situation is illustrated in Fig. 7.2 of Sect. 7.1. The comparison of the values O_k and E_k can help in evaluating the actual degree of symmetry of the dice.

Example 11.2. Normal distribution. A physical quantity is measured N times, and the values are represented in a histogram with $\mathcal{N}$ bins; the height of the kth bin is proportional to the number O_k of observed values. One hypothesizes that the histogram is a finite sample of a normal distribution, with probability density $f(x)$; each bin of the histogram can be compared with a portion of the area under the curve $f(x)$, corresponding to a probability $\mathcal{P}_k$. The expected value for the kth bin is $E_k = N\mathcal{P}_k$. The situation is illustrated in Fig. 7.3 of Sect. 7.1 and Fig. 6.20 of Sect. 6.6.

Example 11.3. Functional relation $Y = \phi(X)$. $\mathcal{N}$ pairs of values (x_k, y_k) are measured; let us suppose that the uncertainties δx_k are negligible with respect to the uncertainties δy_k. One hypothesizes that the measurements can be interpreted by a functional relation $Y = \phi(X)$. For each measured value x_k, one can compare the observed value $O_k = y_k$ with the expected value $E_k = \phi(x_k)$. The situation is illustrated in Fig. 10.3 of Sect. 10.3 for a linear function.

The three examples refer to three rather different situations, in all of which, however, the comparison between theory and experiment can be reduced to the comparison between a set of observed values O_k and a set of corresponding expected values E_k.

The starting point of the chi square test is the expression of the global discrepancy between theory and experiment by the sum of the squared differences between observed and expected values,

$$\sum_{k=1}^{\mathcal{N}} \frac{(O_k - E_k)^2}{(\delta O_k)^2} \ . \tag{11.1}$$

In (11.1), δO_k is the uncertainty of the observed value O_k. The discrepancy between each observed value and the corresponding expected value is thus weighted by the inverse of the uncertainty of the observed value.

11.2 Definition of Chi Square

In order to formally introduce the quantity chi square, it is necessary to discuss the nature of the uncertainties δO_k appearing in (11.1).

In Examples 11.1 and 11.2, the discrete samplings of two distributions have been described in terms of observed values O_k for each value of a random variable K. Each observed value O_k is a discrete random variable, obeying the binomial (Sect. 6.1) or Poisson (Sect. 6.4) limiting distribution, depending on the procedure of measurement (constant N or constant time, respectively). The expected value of O_k is the mean of the binomial or Poisson limiting distribution, $E_k = \mathbf{m}[O_k]$; for the Poisson distribution, mean and variance coincide, so that $E_k = \mathbf{m}[O_k] = \mathbf{D}[O_k]$.

11.2 Definition of Chi Square

It was shown in Sect. 6.6 that both binomial and Poisson distributions can be approximated by a normal distribution if their means are sufficiently large. For the present application, this approximation can be considered valid when E_k is of the order of 5 or larger. The uncertainties $(\delta O_k)^2$ of (11.1) can then be considered as variances σ_k^2 of normal distributions.

In Example 11.3, if the experimental uncertainties are due to random fluctuations, then the uncertainties $(\delta O_k)^2$ of (11.1) can be considered as the variances σ_k^2 of the corresponding normal distributions of sample means.

When the uncertainties δO_k can be expressed as variances of suitable normal distributions, then the sum in (11.1) is referred to as *chi square*:

$$\chi^2 = \sum_{k=1}^{\mathcal{N}} \frac{(O_k - E_k)^2}{\sigma_k^2}. \tag{11.2}$$

The values O_k are random variables, therefore the sum χ^2 is a random variable as well.

Approximate Interpretation of the Chi Square

A first rough interpretation of the chi square can be based on the following arguments. Even if the theory were absolutely true, one would expect a difference between expected and observed values, because of the unavoidable presence of random fluctuations. The difference $O_k - E_k$ would be a random variable, centered on the zero value; the extent of the fluctuations of O_k with respect to E_k would be measured by the uncertainty $\delta O_k = \sigma_k$ of the observed value. Otherwise stated, one would expect that, for each term in the sum (11.2),

$$\frac{(O_k - E_k)^2}{\sigma_k^2} \simeq 1, \tag{11.3}$$

so that

$$\chi^2 = \sum_{k=1}^{\mathcal{N}} \frac{(O_k - E_k)^2}{\sigma_k^2} \simeq \mathcal{N}. \tag{11.4}$$

Let us now suppose that, in a real case, the value of χ^2 has been evaluated through (11.2). An approximate interpretation of the χ^2 value can be made as follows.

(a) A value of χ^2 of the order of $\mathcal{N}$ or smaller indicates that the theory is probably (but not necessarily) correct.
(b) A value of χ^2 much smaller than $\mathcal{N}$ generally indicates that the uncertainties σ_k have been overestimated.
(c) A value of χ^2 much larger than $\mathcal{N}$ generally indicates that the discrepancies between observed and expected values cannot be solely attributed to random fluctuations, and the theory is probably incorrect; however, the

large value of χ^2 could also be due to an underevaluation of the uncertainties σ_k.

The three cases listed above are schematically illustrated in Fig. 11.1, for the case of a linear fit to experimental data points. A rough evaluation of the value of χ^2 can often be made simply by visual inspection of the graph, by comparing, for each point, the discrepancy theory-experiment with the uncertainty bars.

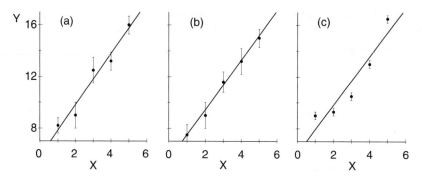

Fig. 11.1. The three graphs show three different situations relative to a linear fit to a set of data points: $\chi^2 \simeq \mathcal{N}$ (left), $\chi^2 \ll \mathcal{N}$ (center), $\chi^2 \gg \mathcal{N}$ (right).

Chi Square and Degrees of Freedom

Let us now try to gain a deeper understanding of the interpretation of the χ^2. The expectation $\chi^2 \simeq \mathcal{N}$ of (11.4) is justified only if the $\mathcal{N}$ terms of the sum are mutually independent. This is the case when the theory is completely independent of experiment. In most cases of interest, however, the theory is to some extent determined by experimental results. As a consequence, not all the $(O_k - E_k)$ terms in (11.2) are independent; once a number ν of $(O_k - E_k)$ terms are known, the remaining $c = \mathcal{N} - \nu$ terms are univocally determined, and lose any random character. The topic can be better clarified by reconsidering the previous examples 11.1 through 11.3.

Example 11.4. Binomial distribution (Example 11.1). Once the characteristics of the phenomenon are determined, the parameters n and p are known, so that the binomial distribution $\mathcal{P}_{np}$ is completely known. The expected values E_k depend, however, also on the sample size: $E_k = N \mathcal{P}_{np}$. Once $\mathcal{N} - 1$ values O_k are known, the $\mathcal{N}$th value is univocally determined by the constraint $\sum O_k = \sum E_k = N$. The sum in (11.2) thus contains $\nu = \mathcal{N} - 1$ independent terms.

Example 11.5. Normal distribution (Example 11.2). When comparing an experimental histogram with a limiting normal distribution, the constraint $\sum O_k = \sum E_k = N$ has again to be taken into account. Moreover, in this case the parameters m and σ of the limiting distribution are also estimated from the corresponding sample parameters; the two equations $m = m^*$ and $\sigma = [N/(N-1)]^{1/2}\sigma^*$ represent two further links between the values O_k and the values E_k: $\sum x_k O_k = \sum x_k E_k$ and $N\sum(x_k - m^*)^2 O_k = (N-1)\sum(x_k - m)^2 E_k$. Globally, there are $c = 3$ constraints, and the sum in (11.2) contains $\nu = \mathcal{N} - 3$ independent terms.

Example 11.6. Functional relation $Y = \phi(X)$ (Example 11.3). Let us first focus attention on a linear relation $Y = A + BX$. If the parameters A and B are estimated from the experimental points by linear regression (Sect. 10.3), then (10.26) and (10.27) represent two links between the values $O_k = y_k$ and the values $E_k = \phi(x_k)$. There are thus $c = 2$ constraints, and the sum in (11.2) contains $\nu = \mathcal{N} - 2$ independent terms. For any function $\phi(X)$, the number of constraints is equal to the number of parameters of the function $\phi(X)$ that are determined from experimental data.

In general, the observed values O_k and the expected values E_k are connected by a number c of independent equations, named constraints. The number of independent terms in the sum in (11.2) is thus $\nu = \mathcal{N} - c$, and is named the number of *degrees of freedom*. Only ν terms, out of the $\mathcal{N}$ terms of the sum, have random character. It is thus reasonable to expect $\chi^2 \simeq \nu$ instead of $\chi^2 \simeq \mathcal{N}$. Obviously, the chi square test is meaningful only if the sum in (11.2) contains at least one random term, say if $\mathcal{N} > c$, so that $\nu > 0$.

Let us consider again the previous examples.

Example 11.7. Binomial distribution (Examples 11.1 and 11.4). For the binomial distribution $\mathcal{P}_{n,p}$ there is only one constraint, $c = 1$. The condition $\nu > 0$, say $\mathcal{N} - c = \nu > 0$, is always fulfilled. In fact, also for $n = 1$, the random variable K has $\mathcal{N} = 2$ possible values, $k_1 = 0$ and $k_2 = 1$. One expects

$$\chi^2 = \sum_{k=1}^{\mathcal{N}} \frac{(O_k - N\mathcal{P}_{np})^2}{\sigma_k^2} \simeq \mathcal{N} - 1. \tag{11.5}$$

Example 11.8. Normal distribution (Examples 11.2 and 11.5). For the normal distribution, there are three constraints. In order that $\nu > 0$, the number $\mathcal{N}$ of bins of the experimental histogram cannot be less than four. One thus expects

$$\chi^2 = \sum_{k=1}^{\mathcal{N}} \frac{(O_k - E_k)^2}{\sigma_k^2} \simeq \mathcal{N} - 3. \tag{11.6}$$

Example 11.9. Functional relation $Y = \phi(X)$ (Examples 11.3 and 11.6). The number of constraints is equal to the number of parameters of the function $\phi(x)$ determined from experiment. In order that $\nu > 0$, the number $\mathcal{N}$ of experimental points has to be larger than the number of parameters.

For the *direct proportionality* $Y = BX$, there is one parameter B; if there were only one experimental point, the best fitting straight line would certainly include it, independently of uncertainty; for the chi square test to be meaningful, the number $\mathcal{N}$ of experimental points should be at least two, and one expects

$$\chi^2 = \sum_{k=1}^{\mathcal{N}} \frac{(y_k - Bx_k)^2}{\sigma_k^2} \simeq \mathcal{N} - 1. \qquad (11.7)$$

For a *linear relation* $Y = A+BX$, there are two parameters A, B; if there were only two experimental points, the best fitting straight line would certainly include them, independently of their uncertainties; for the chi square test to be meaningful, the number $\mathcal{N}$ of experimental points should be at least three, and one expects

$$\chi^2 = \sum_{k=1}^{\mathcal{N}} \frac{(y_k - A - Bx_k)^2}{\sigma_k^2} \simeq \mathcal{N} - 2. \qquad (11.8)$$

The discrepancy between theory and experiment has a random character, and χ^2 is a random variable. A thorough probabilistic interpretation of the meaning of χ^2 is given in Sect. 11.3. Let us give here some rules of thumb for interpreting the symbol $\simeq$ appearing in (11.5) through (11.8).

(a) If $\chi^2 \simeq \mathcal{N} - c = \nu$ (and if one is confident that the uncertainties σ_k^2 have been correctly evaluated), then it is reasonable to assume that the theory is consistent with experimental results.
(b) If $\chi^2 \ll \mathcal{N} - c = \nu$, then it is still reasonable to assume that the theory is consistent with experimental results; there is, however, a nonnegligible probability that the uncertainties σ_k^2 have been overestimated.
(c) If $\chi^2 \gg \mathcal{N} - c = \nu$, then it is probable that theory and experimental results are inconsistent, unless the uncertainties σ_k^2 have been strongly underestimated.

11.3 The Chi Square Distribution

Once a theoretical hypothesis has been made, the expected values E_k are determined; the experimental values O_k are instead random variables. As a consequence, χ^2 is also a random variable. If the probability distribution of the random variable χ^2 is known, one can evaluate the probability of obtaining χ^2 values larger or smaller than the value actually obtained by experiment. This procedure leads to a quantitative probabilistic interpretation of the χ^2 values.

Definition of the Chi Square Distribution

To find out the χ^2 distribution, it is worth observing that all terms in the sum in (11.2) have the same form, independently of the nature of the problem: the variable O_k is normally distributed around the value E_k, with variance σ^2. Each term of the sum in (11.2) corresponds then to the standard normal variable (Sect. 6.5),

$$Z = \frac{O_k - E_k}{\sigma_k}, \qquad (11.9)$$

whose mean is zero and whose standard deviation is one. The χ^2 is thus the sum of $\mathcal{N}$ standard normal variables,

$$\chi^2 = \sum_{k=1}^{\mathcal{N}} Z^2. \qquad (11.10)$$

The $\mathcal{N}$ terms of the sum (11.10) are generally not independent. The number of independent terms is given by the number ν of degrees of freedom, so that the sum in (11.10) contains $\nu \leq \mathcal{N}$ independent random variables Z^2, all characterized by the same distribution law.

The distribution law of the random variable χ^2 defined by (11.10) can be well approximated by the distribution law of a quantity χ^2 defined as the sum of ν independent variables Z^2:

$$\chi^2 = \sum_{k=1}^{\nu} Z^2. \qquad (11.11)$$

Strictly speaking, in probability theory the χ^2 is the sum of independent Z^2 variables, Z being the standard normal variable, as in (11.11). Our definition (11.10) is a widely accepted extension of terminology. The chi square distribution is the distribution of the random variable χ^2 defined by (11.11) as the sum of independent Z^2 variables.

The chi square distribution depends on the number ν of terms in the sum (11.11). There is thus a different distribution $G_\nu(\chi^2)$ for each different value of the number ν of degrees of freedom.

Analytical Expression of the Chi Square Distribution

It is relatively easy to evaluate the chi square distribution for one degree of freedom, $\nu = 1$; in this case

$$\chi^2 = z^2, \quad \text{where } f(z) = \frac{1}{\sqrt{2\pi}} \exp\left[-z^2/2\right]. \qquad (11.12)$$

To calculate the probability density $G(z^2) \equiv G_1(\chi^2)$, one can notice that to each value z^2 of the variable Z^2 correspond two values of Z: $+z$ and $-z$. The probability that the value of Z^2 belongs to an interval $d(z^2)$ is

$$G(z^2)\,\mathrm{d}(z^2) = 2\,f(z)\,\mathrm{d}z$$
$$= 2\,f(z)\,\frac{\mathrm{d}z}{\mathrm{d}(z^2)}\,\mathrm{d}(z^2) = 2\,f(z)\,\frac{1}{2z}\,\mathrm{d}(z^2)\,, \tag{11.13}$$

so that
$$G(z^2) = \frac{f(z)}{z} = \frac{1}{z\sqrt{2\pi}}\,\exp\!\left[-z^2/2\right]\,, \tag{11.14}$$

say
$$G_1(\chi^2) = 1\sqrt{2\pi\chi^2}\,\exp\!\left[-\chi^2/2\right]\,. \tag{11.15}$$

The function $G_1(\chi^2)$ diverges for $\chi^2 \to 0$ (Fig. 11.2, left); one can always verify that it is normalized to one.

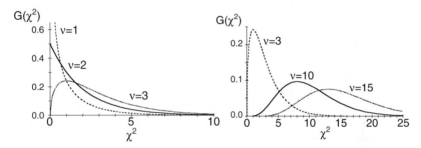

Fig. 11.2. Chi square distributions $G_\nu(\chi^2)$ for different degrees of freedom. Notice that the horizontal and vertical scales of the left- and right-handside graphs are different.

For two degrees of freedom ($\nu = 2$), one can demonstrate that the chi square distribution is (Fig. 11.2, left)
$$G_2(\chi^2) = G(Z^2 + Z^2) = 0.5\,\exp\!\left[-\chi^2/2\right]\,. \tag{11.16}$$

The chi square distribution for a number of degrees of freedom ν is given by the expression
$$G_\nu(\chi^2) = \frac{1}{2^{\nu/2}\,\Gamma(\nu/2)}\,(\chi^2)^{\nu/2-1}\,\exp\!\left[-\chi^2/2\right]\,, \tag{11.17}$$

where the Γ function is defined as follows,

For n integer $\Gamma(n+1) = n!$
For n half-integer $\Gamma(n+1/2) = (2n-1)\times(2n-3)\cdots 5\times 3\times\sqrt{\pi}/2^n$.

Some chi square distributions for different degrees of freedom ν are shown in Fig. 11.2.

Mean and Variance

The mean of the chi square distribution with ν degrees of freedom is:

$$\langle \chi^2 \rangle = \left\langle \sum_{k=1}^{\nu} Z^2 \right\rangle = \sum_{k=1}^{\nu} \langle Z^2 \rangle = \nu \langle Z^2 \rangle. \tag{11.18}$$

To calculate $\langle Z^2 \rangle$, remember that Z is the standard normal variable (11.19), so that

$$\langle Z^2 \rangle = \left\langle \frac{(O_k - E_k)^2}{\sigma_k^2} \right\rangle = \frac{\langle (O_k - E_k)^2 \rangle}{\sigma_k^2} = \frac{\sigma_k^2}{\sigma_k^2} = 1. \tag{11.19}$$

In conclusion, the mean of the chi square distribution is equal to the number of degrees of freedom:

$$\langle \chi^2 \rangle = \nu. \tag{11.20}$$

This conclusion justifies the approximate interpretation introduced at the end of Sect. 11.2, based on the expectation that the experimental value of χ^2 be equal to the number of degrees of freedom.

One can also demonstrate that the variance of the chi square distribution is twice the number of degrees of freedom, $\mathbf{D}[\chi^2] = 2\nu$.

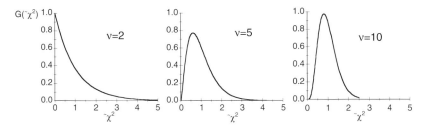

Fig. 11.3. Distributions of the reduced chi square $G_\nu(\tilde{\chi}^2)$ for $\nu = 2, 5, 10$.

Reduced Chi Square

The reduced chi square $\tilde{\chi}^2$ is frequently utilized instead of the chi square χ^2. The *reduced chi square* $\tilde{\chi}^2$ is defined as the ratio of the chi square χ^2 to the number ν of degrees of freedom

$$\tilde{\chi}^2 = \chi^2/\nu. \tag{11.21}$$

The expectation $\chi^2 \simeq \nu$ corresponds then to the expectation $\tilde{\chi}^2 \simeq 1$.

It is also possible to calculate the distributions $G_\nu(\tilde{\chi}^2)$ of the reduced chi square. Some distributions are shown in Fig. 11.3. One can appreciate that the distributions of the reduced chi square become narrower when ν increases. Actually, the mean of the distributions is always one, and the variance decreases as ν increases.

11.4 Interpretation of the Chi Square

The knowledge of the chi square distributions $G_\nu(\chi^2)$ allows a probabilistic interpretation of the value of χ^2 obtained in a given experiment.

Let us suppose that an experiment has led to $\mathcal{N}$ experimental values O_k, whereas a theoretical hypothesis leads to $\mathcal{N}$ corresponding expected values E_k, the number of degrees of freedom being ν. The chi square, calculated according to (11.2), is labeled here as χ_o^2, the index "o" standing for "observed",

$$\chi_o^2 = \sum_{k=1}^{\mathcal{N}} \frac{(O_k - E_k)^2}{\sigma_k^2}, \qquad (11.22)$$

to distinguish it from the generic value χ^2.

The problem is now to evaluate whether the observed discrepancy, measured by the value χ_o^2, can solely be attributed to random fluctuations of the observed values O_k with respect to the expected values E_k, or is due to the inadequacy of the theoretical hypothesis.

If the theoretical hypothesis were correct, the expected values of χ^2 due to random fluctuations would be distributed according to $G_\nu(\chi^2)$.

The integral

$$\int_0^{\chi_o^2} G_\nu(\chi^2) \, \mathrm{d}\chi^2 = \mathcal{P}\left[\chi^2 < \chi_o^2\right], \qquad (11.23)$$

where the upper limit is the observed value χ_o^2, is the probability of obtaining, merely because of random fluctuations, a value *smaller* than the value χ_o^2 actually observed. By converse, the integral

$$\int_{\chi_o^2}^{\infty} G_\nu(\chi^2) \, \mathrm{d}\chi^2 = \mathcal{P}\left[\chi^2 > \chi_o^2\right] \qquad (11.24)$$

is the probability of obtaining a value *larger* than the value χ_o^2 actually observed.

Analogous expressions are obtained for the reduced chi square,

$$\int_0^{\tilde{\chi}_o^2} G_\nu(\tilde{\chi}^2) \, \mathrm{d}\tilde{\chi}^2 = \mathcal{P}\left[\tilde{\chi}^2 < \tilde{\chi}_o^2\right], \qquad (11.25)$$

$$\int_{\tilde{\chi}_o^2}^{\infty} G_\nu(\tilde{\chi}^2) \, \mathrm{d}\tilde{\chi}^2 = \mathcal{P}\left[\tilde{\chi}^2 > \tilde{\chi}_o^2\right]. \qquad (11.26)$$

Calculating the integrals (11.23) through (11.26) is far from trivial; the corresponding values are tabulated in various equivalent forms; see, for example, Appendix C.5.

It is worth noting that the comparison between the observed values χ_o^2 or $\tilde{\chi}_o^2$ and the tabulated values of the integrals (11.23) through (11.26) does not allow a deterministic decision about the soundness of a theoretical hypothesis; it can only lead to a probabilistic evaluation:

(a) If the probability $\mathcal{P}\left[\tilde{\chi}^2 > \tilde{\chi}_0^2\right]$ is considered large, then it is reasonable to assume that the theoretical hypothesis is consistent with experimental data.
(b) If the probability $\mathcal{P}\left[\tilde{\chi}^2 > \tilde{\chi}_0^2\right]$ is considered small, then it is reasonable to doubt the soundness of the theoretical hypothesis.

By convention, if $\mathcal{P}\left[\tilde{\chi}^2 > \tilde{\chi}_0^2\right] \simeq 0.05$, one considers that there is *significant disagreement* between theory and experiment; if $\mathcal{P}\left[\tilde{\chi}^2 > \tilde{\chi}_0^2\right] \simeq 0.01$ the disagreement is said to be *highly significant*.

Anyway, it should be borne in mind that the reliability of the chi square test often depends on the correct evaluation of the experimental uncertainties σ_k.

Problems

11.1. Attempt a rough evaluation of the chi squares for the three graphs represented in Fig. 11.1, by using a common ruler. Compare the chi square values with the number of degrees of freedom (remember that the best fitting straight line has been obtained by linear regression).

11.2. Consider again the three graphs of Fig. 11.1. The coordinates of the experimental points, x_k and $y_k = O_k$, the uncertainties $\delta y_k = \sigma_k$ and the $y_k = E_k$ values of the best fitting straight line $A + BX$ for graph (a) are (in arbitrary units):

X	1	2	3	4	5
Y	8.2	9.0	12.5	13.2	16.0
δY	0.6	1.0	1.0	0.7	0.7
$A + BX$	7.8	9.8	11.8	13.8	15.7

The numerical values for graph (b) are:

X	1	2	3	4	5
Y	7.5	9.0	11.6	13.2	15.0
δY	0.8	1.0	0.8	1.0	0.7
$A + BX$	7.5	9.4	11.3	13.3	15.2

The numerical values for graph (c) are:

X	1	2	3	4	5
Y	9.0	9.3	10.5	13.0	16.5
δY	0.3	0.3	0.3	0.3	0.3
$A + BX$	7.9	9.8	11.7	13.5	15.4

For each of the three sets of values, calculate the chi square (11.2). Compare the χ^2 values with the rough values obtained in Problem 11.1 by the graphical procedure.

Compare the χ^2 values with the number of degrees of freedom, $\nu = 3$, and evaluate the soundness of the hypothesis of linearity for the three different cases.

For each of the three graphs, evaluate the probability that the expected value of χ^2 is larger than the observed value, making use of the tables of Appendix C.5.

Part IV

Appendices

A Presentation of Experimental Data

A.1 Significant Digits and Rounding

In science and technology, one sometimes deals with exact numerical values.

Example A.1. The sine function has an exact value when the argument is $\pi/6$: $\sin(\pi/6) = 0.5$.

Example A.2. The velocity of light c, expressed in $\mathrm{m\,s^{-1}}$, has by convention the exact value 299 792 458.

More often, one deals with approximate numerical values.

Example A.3. The value of the cosine function when the argument is $\pi/6$ can be expressed only in approximate ways, depending on the required accuracy; for example, $\cos(\pi/6) \simeq 0.866$, or $\cos(\pi/6) \simeq 0.8660254$.

Example A.4. The value of the velocity of light is often expressed in approximate form, $c \simeq 3 \times 10^8\,\mathrm{m\,s^{-1}}$.

Example A.5. The measure of a physical quantity is always an approximate value, the extent of the approximation being connected to the uncertainty.

Significant Digits

The number of significant digits of a numerical value is obtained by counting the digits from left to right, starting from the first nonzero digit. The zeroes to the left of significant digits have only positional character. For example:

> The number 25.04 has 4 significant digits: 2 5 0 4
> The number 0.0037 has 2 significant digits: 3 7
> The number 0.50 has 2 significant digits: 5 0

Counting the significant digits is not unambiguous when dealing with integer values terminating with at least one zero, such as 350 or 47000. In these cases, it is not evident whether the zeroes are actually significant or only positional. The scientific notation avoids ambiguities.

Example A.6. Let us consider the value 2700. Using scientific notation, one can specify the number of significant zeroes:

$$2700 = \begin{array}{l} 2.7 \times 10^3 \text{ (2 significant digits)} \\ 2.70 \times 10^3 \text{ (3 significant digits)} \\ 2.700 \times 10^3 \text{ (4 significant digits)} \end{array}$$

The following nomenclature is used for significant digits.

– The first digit on the left is the *most significant digit* (MSD).
– The last digit on the right is the *least significant digit* (LSD).

Sometimes, the measurement uncertainty is implicitly implied in the number of significant digits. For example:

$$X = 2.47 \text{ m} \quad \text{instead } X = (2.47 \pm 0.005) \text{ m}$$
$$X = 2.470 \text{ m} \quad \text{instead } X = (2.470 \pm 0.0005) \text{ m}$$

The implicit expression of uncertainty should be avoided, because in some cases it can be equivocal.

Rules for Rounding off Numerical Values

When the number of digits of a numerical value has to be reduced, the remaining least significant digit has to be rounded off, according to the following rules.

1. If the most significant digit to be eliminated is 0, 1, 2, 3, 4, then the remaining least significant digit is left unaltered. For example, 12.34 ≃ 12.3.
2. If the most significant digit to be eliminated is 6, 7, 8, 9, or 5 followed by at least a nonzero digit, then the remaining least significant digit is incremented by one. Examples: 12.36 ≃ 12.4; 12.355 ≃ 12.4.
3. If the most significant digit is 5 followed only by zeroes, then the remaining least significant digit is left unaltered if it is even; it is incremented by one if it is odd. Examples: 12.45 ≃ 12.4, 12.35 ≃ 12.4.

Rounding off the Results of Calculations

When calculations are performed on approximate numerical values, not all digits of the result are necessarily significant. In this case, the result has to be rounded off, so as to maintain only the significant digits. Although no exact prescriptions exist, some rules of thumb can be given.

For *additions and subtractions* of approximate numbers: the digits of the sum or difference are not significant to the right of the position corresponding to the leftmost of the least significant digits of the starting terms.

Example A.7. Let us sum up the three approximate numbers: 2.456, 0.5, 3.35; because the second term has no significant digits beyond the first decimal position, the sum is rounded to the first decimal position:

```
            2.456   +
            0.5     +
            3.35    =
            ─────
            6.306   ⟶   6.3
```

Example A.8. Let us calculate the average of the three approximate numbers: `19.90`, `19.92`, `19.95`. The average value `19.923333`, obtained by a calculator, has to be rounded to `19.92`.

For *multiplications and divisions* of approximate numbers: it is reasonable to round off the result to n, or sometimes $n+1$ significant digits, where n is the smallest between the numbers of significant digits of the factors.

The *square roots* of approximate numbers are generally rounded off to the same number of significant digits of the radicand.

Example A.9. The product of the approximate numbers `6.83` and `72` is evaluated by a calculator. The result `491.76` is rounded off to two significant digits, 4.9×10^2.

Example A.10. The approximate number `83.642` is divided by the approximate number `72`. The results `1.1616944` can be rounded off to two significant digits, `1.2`, but in this case it can be preferable to maintain also the third digit, `1.16`.

Example A.11. The square root of `30.74` is evaluated by a calculator. The result $\sqrt{30.74} = $ `5.5443665` is rounded off to `5.544`.

Example A.12. The tangent of an angle of $27°$ is calculated as the ratio between the sine and cosine values, approximate to two significant digits: $\sin(27°) \simeq$ `0.45` and $\cos(27°) \simeq$ `0.89`; the value `0.505618`, obtained by a calculator, is approximated to `0.51` or `0.506`. The direct calculation of the tangent by the same calculator would give the value $\tan(27°) \simeq$ `0.5095398`.

Significant Digits and Measurement Uncertainty

The number of significant digits of a measure is determined by the extent of its uncertainty. Rounding techniques are generally used for the expression of uncertainty. Two rules of thumb should be taken into account.

(a) The uncertainty δX should be expressed by no more than two significant digits, and sometimes one significant digit is sufficient.
(b) When a measurement result is expressed as $X_0 \pm \delta X$, the least significant digit of X_0 should be of the same order of magnitude as the least significant digit of the uncertainty.

For measurements affected by random fluctuations (Sect. 4.3), the central value X_0 and the uncertainty δX are evaluated from the experimental data as

$$X_0 = \frac{1}{N} \sum_i x_i, \qquad \delta X = \sqrt{\frac{1}{N(N-1)} \sum_i (x_i - m^*)^2}. \qquad \text{(A.1)}$$

The values X_0 and δX obtained by (A.1) generally include nonsignificant digits, and should be rounded off according to the aforementioned rules of thumb.

Example A.13. The period T of a pendulum is measured $N = 6$ times, and the following six values are obtained (expressed in seconds):
$$2.15,\ 2.14,\ 2.17,\ 2.15,\ 2.16,\ 2.17.$$
By (A.1), one obtains the approximate values
$$T = 2.1566667\,\text{s}, \qquad \delta T = 0.004944\,\text{s},$$
that should be rounded off as
$$T = 2.157\,\text{s}, \qquad \delta T = 0.005\,\text{s}.$$

Example A.14. The uncertainty of a length value is calculated as $\delta X = 0.015$ mm. In this case, it is not convenient to round off the value to one significant digit, $\delta X = 0.01$ mm, because a nonnegligible contribution to uncertainty would be eliminated.

In indirect measurements, both central value and uncertainty are obtained from calculations performed on directly measured values (Chap. 8). Also in indirect measurements, it can be necessary to round off the results.

Example A.15. The side a of a square is directly measured, obtaining the value $a = (23 \pm 0.5)$ mm. The length of the diagonal d is calculated as $d = a\sqrt{2} = 32.526912 \pm 0.707106$ mm. The value has to be rounded off to $d = 32.5 \pm 0.7$ mm.

Example A.16. The acceleration of gravity g is evaluated from the directly measured values of length ℓ and period T of a pendulum: $g = (2\pi/T)^2\,\ell$. One obtains: $g_0 = 9.801357\,\text{m\,s}^{-2}$; $\delta g = 0.023794\,\text{m\,s}^{-2}$. The uncertainty is rounded off to $\delta g = 0.024\,\text{m\,s}^{-2}$. The value of g_0 is rounded off accordingly, so that finally: $g = g_0 \pm \delta g = (9.801 \pm 0.024)\,\text{m\,s}^{-2}$.

A.2 Tables

A table is an effective and synthetic method for presenting correlated entities. A typical example is the table connecting names and symbols of lower- and upper-case Greek letters (Appendix C.1). Here we are mainly interested in mathematical and physical tables.

Mathematical Tables

Some mathematical functions cannot be evaluated by simple and fast analytical algorithms: their calculation is based on laborious procedures that are convenient to perform once for all. In such cases, tables are frequently used, in which the values y_i of the function are shown in correspondence with selected values x_i of the independent variable. In general, the significant digits appearing in the table indicate the precision of the calculation.

Example A.17. Before the advent of calculators in the 1970s, the values of trigonometric functions, exponentials, and logarithms were listed in published tables, whose layout was like that of Table A.1.

Table A.1. An example of a table of trigonometric functions.

Angle (deg)	Angle (rad)	Sin	Cos	Tan
0°	0.000	0.000	1.000	0.000
2°	0.035	0.035	0.999	0.035
4°	0.070	0.070	0.998	0.070
6°	0.105	0.105	0.995	0.105
8°	0.140	0.139	0.990	0.141
...	...	...	...	...

Example A.18. The integrals of several functions relevant for the theory of probability, such as the normal distribution, the Student distribution, and the chi square distribution, are listed in some tables of Appendix C.

Physical Tables

Tables are frequently utilized to compare and possibly correlate the values of two or more physical quantities. The values of each quantity are listed in a column of the table. At the top of the column, the name of the physical quantity and its unit are written. For example, Table A.2 lists the measures of the period of a pendulum as a function of its length.

The values of physical quantities must always be accompanied by the corresponding uncertainties, for example, in the standard form $X_0 \pm \delta X$ (Table A.2, left-handside). If the uncertainty is the same for all values, it can be given once at the top of the corresponding column (Table A.2, right-handside).

Linear Interpolation

Let us consider a table containing the values of two quantities X and Y. Only a finite number of pairs (x_i, y_i) can be actually listed. Sometimes one wants

Table A.2. Tabular representation of the values of the period of a pendulum measured as a function of the length. The uncertainties have been quoted for each value in the left-hand side, only once for all in the right-hand side.

Length (cm)	Period (s)	Length (±0.05 cm)	Period (±0.005 s)
20 ± 0.05	0.89 ± 0.005	20	0.89
40 ± 0.05	1.26 ± 0.005	40	1.26
60 ± 0.05	1.55 ± 0.005	60	1.55
80 ± 0.05	1.79 ± 0.005	80	1.79

to know the value $\tilde{y}$ corresponding to a value $\tilde{x}$ that does not appear in the table, but is included between two listed values: $x_1 < \tilde{x} < x_2$. The problem can be easily solved if the relation between X and Y can be considered as linear, at least locally (say within the interval $x_1 - x_2$); one can then linearly interpolate

$$\frac{\tilde{y} - y_1}{\tilde{x} - x_1} = \frac{y_2 - y_1}{x_2 - x_1}, \quad \text{so that} \quad \tilde{y} = y_1 + \frac{y_2 - y_1}{x_2 - x_1}(\tilde{x} - x_1). \quad (\text{A.2})$$

If the relation cannot be considered linear, one must rely on more complex interpolation procedures, typically based on polynomials of order higher than one.

A.3 Graphs

Graphs allow a synthetic and suggestive representation of the values of two or more physical quantities. Only two-dimensional graphs (x, y), with orthogonal axes, are considered here, the extension to more complex situations being quite trivial. Typical applications of graphs in physics are:

(a) The visualization of the behavior of a function $y = \phi(x)$ in a finite region of its domain (Fig. A.1, left)
(b) The search for a correlation between the measured values of two quantities X and Y, and possibly of a suitable functional dependence $y = \phi(x)$ (Fig. A.1, right)

General Considerations

Some simple rules can help to enhance the readability and effectiveness of graphs.

1. The independent and dependent variables should be represented on the horizontal and vertical axes, respectively.

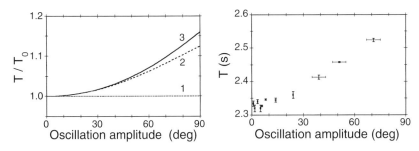

Fig. A.1. The period T of a pendulum depends on the oscillation amplitude θ_0 according to: $T = T_0 \left[1 + (1/4)\sin^2(\theta_0/2) + (9/64)\sin^4(\theta_0/2) + \cdots\right]$, where $T_0 = 2\pi(\ell/g)^{1/2}$. The graph on the left shows the calculated ratio T/T_0: curves 1, 2, and 3 refer to the approximations of first-, second- and third-order, respectively. The graph on the right shows several experimental values measured with a pendulum of length $\ell = 136.9\,\mathrm{cm}$.

2. When the graph refers to physical quantities, it is necessary to indicate both the name and the unit corresponding to each axis, as in Fig. A.1.
3. The scales should be chosen so that the coordinates of the points can be easily determined. The ticks should then correspond to equally spaced and rounded numerical values, such as 0, 2, 4, 6, or 0, 10, 20, 30. One should avoid, if possible, nonrounded values, such as 1.2, 2.4, 3.6, 4.8. One should always avoid nonequally spaced values, such as 1.2, 2.35, 2.78, 3.5, even if corresponding to measured values.
4. When plotting experimental values, the uncertainties are represented by segments (error bars): horizontal and vertical for the quantities represented on the horizontal and vertical axes, respectively (Fig. A.1, right).
5. The scales are not necessarily linear; they can often be suitably chosen so that the plotted points align along a straight line. The most common types of scales are described below.

Linear Scales

Graphs with linear scales are useful to visually check the linearity of the relation between two quantities x and y, such as $y = Bx$ (Fig. A.2, left), or $y = A + Bx$ (Fig. A.2, right).

Once the linear dependence of y on x has been recognized, the parameters A and B and the corresponding uncertainties δA and δB can be evaluated

(a) Graphically, by drawing the straight lines of minimum and maximum slope compatible with the error bars
(b) Analytically, by the linear regression procedure of Sect. 10.3

The soundness of the hypothesis of linearity can be checked a posteriori by the chi square test introduced in Chap. 11.

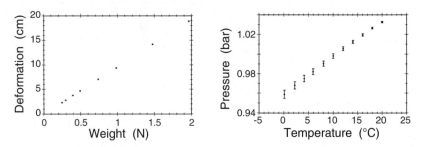

Fig. A.2. Graphs with linear scales. Left: Deformation of a spring measured as a function of the applied force (the error bars are smaller than the symbol size). Right: Pressure of a gas measured as a function of temperature, at constant volume (the vertical error bars take into account the uncertainties of both temperature and pressure).

Semilog Graphs

Exponential relations between two quantities x and y, such as

$$y = a\,e^{bx} \tag{A.3}$$

(Fig. A.3, left), can be linearized by representing on the vertical axis the natural logarithm, $Y = \ln y$ (Fig. A.3, center). The relation between x and $Y = \ln y$ is linear:

$$\ln y = \ln a + bx\,, \quad \text{say } Y = A + bx\,. \tag{A.4}$$

The semilog graph allows one to visually check if the points (x_i, Y_i) are connected by a linear relation such as (A.4). The parameters $A = \ln a$ and b can again be determined by linear regression (Sect. 10.3).

Alternatively, one can directly plot the original values (x_i, y_i) on semilog paper, or, if computer programs are used, by selecting the option *logarithmic scale* for the vertical axis (Fig. A.3, right).

It is worth remembering that the argument of transcendental functions, such as the logarithms, are dimensionless (Sect. 2.5). In the expression $Y = \ln y$, y stands for the (dimensionless) numerical values of the physical quantity with respect to a given unit, not for the quantity itself.

Log–log Graphs

Power-law relations between two quantities x and y, such as

$$y = a\,x^b \tag{A.5}$$

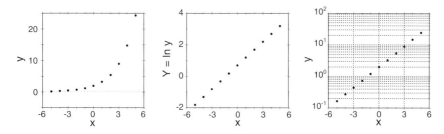

Fig. A.3. Different graphs of points (x_i, y_i) obeying the law $y = a \exp(bx)$, where $a = 2$ and $b = 0.5$. Left: points (x_i, y_i) and linear scales. Center: points $(x_i, \ln y_i)$ and linear scales. Right: points (x_i, y_i) and logarithmic vertical scale.

(Fig. A.4, left), can be linearized by representing $Y = \ln y$ on the vertical scale and $X = \ln x$ on the horizontal scale (Fig. A.4, center). The relation between $X = \ln x$ and $Y = \ln y$ is linear:

$$\ln y = \ln a + b \ln x, \quad \text{say } Y = A + bX. \tag{A.6}$$

The log–log graph allows one to visually check if the points (X_i, Y_i) are connected by a linear relation such as (A.6). The parameters $A = \ln a$ and b can again be determined by linear regression (Sect. 10.3).

Alternatively, one can directly plot the original values (x_i, y_i) on log–log paper, or, if computer programs are used, by selecting the option *logarithmic scale* for both horizontal and vertical axes (Fig. A.4, right).

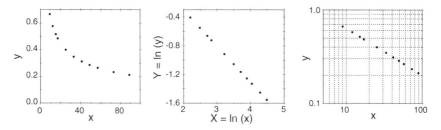

Fig. A.4. Different graphs of points (x_i, y_i) obeying the law $y = a x^b$, where $a = 2$ and $b = -0.5$. Left: points (x_i, y_i) and linear scales. Center: points $(\ln x_i, \ln y_i)$ and linear scales. Right: points (x_i, y_i) and logarithmic horizontal and vertical scales.

Other Scales

In addition to logarithmic scales, other scales can help to linearize particular types of functions. Let us list some examples.

(a) The relation $y = a\sqrt{x}$ can be linearized by representing $\sqrt{x}$ instead of x on the horizontal axis, or y^2 instead of y on the vertical axis. Obviously, because $y = a\sqrt{x} = a\,x^{1/2}$, the relation can be alternatively linearized by a log–log graph.

(b) A relation of inverse proportionality, $xy = K$, can be linearized by plotting y as a function of K/x.

A.4 Histograms

Histograms are graphs where the quantity on the horizontal axis has discrete values, often equally spaced, and the quantity on the vertical axis is represented by the height of a column (bin).

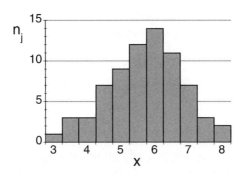

Fig. A.5. Example of histogram. The horizontal axis is divided into intervals of width $\Delta x = 0.5$. If the quantity on the horizontal axis is a physical quantity, the unit has to be specified, for example: ℓ (mm), T (s), and so on.

The $\mathcal{N}$ discrete values on the horizontal axis are labeled by an index j ($j = 1, \ldots, \mathcal{N}$). Often, the discrete values on the horizontal axis correspond to intervals Δx_j of a quantity x, and the quantity on the vertical axis, say the bin height, is an integer n_j (Fig. A.5). If x is a physical quantity, its unit has to be explicitly indicated. If N is the sum over all bins of the n_j values, then

$$\sum_{j=1}^{\mathcal{N}} n_j = N \ . \tag{A.7}$$

Example A.19. The length ℓ of N objects is measured (Experiment E.1 of Appendix E). The dispersion of values can be represented by a histogram, where the bin width corresponds to the resolution $\Delta\ell$ of the measuring instrument.

Example A.20. The period T of a pendulum is measured N times by means of a stopwatch with resolution $0.01\,\mathrm{s}$ (Experiment E.2 of Appendix E). The dispersion of values due to random fluctuations can be represented by a histogram, where the bin width ΔT_j corresponds to the instrument resolution.

Height-Normalized Histograms

In *height-normalized* histograms, the height of each bin is proportional to the ratio

$$p_j^* = n_j/N \,, \tag{A.8}$$

that represents the sample frequency of the jth value. One can easily verify that the sum of the bin heights is always equal to one, independently of the value of N:

$$\sum_{j=1}^{\mathcal{N}} n_j/N = 1 \,. \tag{A.9}$$

The normalization to unit height is necessary when one compares different histograms related to the same quantity x, but with different numbers $N_A, N_B, N_C, \ldots$ of total values.

Area-Normalized Histograms

In *area-normalized* histograms, the height of each bin is proportional to the sample density

$$f_j^* = \frac{n_j}{N \, \Delta X_j} \,. \tag{A.10}$$

One can easily verify that the sum of the bin areas is always equal to one, independently of the value of N:

$$\sum_{j=1}^{\mathcal{N}} f_j^* \, \Delta X_j = 1 \,. \tag{A.11}$$

The normalization to unit area is necessary when one compares different histograms having bins of different widths $\Delta x_A, \Delta x_B, \ldots$, for example, when comparing the results of measurements performed with different instrumental resolutions. Another important application is the comparison of histograms with distributions of continuous random variables (Sect. 4.3).

B Systems of Units

B.1 The International System of Units (SI)

SI: Base Units

The SI is based on seven base quantities. The definition of their units is given below. For each unit, the conference (*Conférence Générale des Poids et Mesures*, CGPM) that defined it, and the year, are given in parentheses.

- **Time.** The second (s) is the duration of 9 192 631 770 periods of the radiation corresponding to the transitions between the two hyperfine levels of the ground state of the cesium 133 atom (13th GCPM, 1967).
- **Length.** The meter (m) is the length of the path travelled by light in vacuum during a time interval of 1/299 792 458 of a second (17th CGPM, 1983).
- **Mass.** The kilogram (kg) is the unit of mass; it is equal to the mass of the international prototype of the kilogram (3rd CGPM, 1901).
- **Electric current.** The ampere (A) is that constant current which, if maintained in two straight parallel conductors of infinite length, of negligible circular cross-section, and placed 1 meter apart in vacuum, would produce between these conductors a force equal to 2×10^{-7} newton per meter of length (9th CGPM, 1948).
- **Thermodynamic temperature.** The kelvin (K) is the fraction 1/273.16 of the thermodynamic temperature of the triple point of water (13th CGPM, 1967).
- **Luminous intensity.** The candela (cd) is the luminous intensity, in a given direction, of a source that emits monochromatic radiation of frequency 540×10^{12} hertz and that has a radiant intensity of 1/683 watt per steradian (16th GCPM, 1979).
- **Amount of substance.** The mole (mol) is the amount of substance of a system that contains as many elementary entities as there are atoms in 0.012 kilogram of carbon 12. When the mole is used, the elementary entities must be specified and may be atoms, molecules, ions, electrons, other particles, or specified groups of such particles (14th CGPM, 1971).

SI: Derived Units

The SI derived units with special names and symbols are listed in Table B.1.

Table B.1. SI derived units with special names.

Quantity	Unit	Symbol	Conversion	Notes
Plane angle	radian	rad	$1\,\text{rad} = 1\,\text{m}\,\text{m}^{-1}$	
Solid angle	steradian	sr	$1\,\text{sr} = 1\,\text{m}^2\,\text{m}^{-2}$	
Frequency	hertz	Hz	$1\,\text{Hz} = 1\,\text{s}^{-1}$	
Force	newton	N	$1\,\text{N} = 1\,\text{m}\,\text{kg}\,\text{s}^{-2}$	
Pressure, stress	pascal	Pa	$1\,\text{Pa} = 1\,\text{N}\,\text{m}^{-2}$	
Work, energy, heat	joule	J	$1\,\text{J} = 1\,\text{N}\,\text{m}$	
Power	watt	W	$1\,\text{W} = 1\,\text{J}\,\text{s}^{-1}$	
Electric charge	coulomb	C	$1\,\text{C} = 1\,\text{A}\,\text{s}$	
Electric potential	volt	V	$1\,\text{V} = 1\,\text{W}\,\text{A}^{-1}$	
Capacitance	farad	F	$1\,\text{F} = 1\,\text{C}\,\text{V}^{-1}$	
Electric resistance	ohm	Ω	$1\,\Omega = 1\,\text{V}\,\text{A}^{-1}$	
Electric conductance	siemens	S	$1\,\text{S} = 1\,\Omega^{-1}$	
Magnetic flux	weber	Wb	$1\,\text{Wb} = 1\,\text{V}\,\text{s}$	
Magnetic flux density	tesla	T	$1\,\text{T} = 1\,\text{Wb}\,\text{m}^{-2}$	
Inductance	henry	H	$1\,\text{H} = 1\,\text{Wb}\,\text{A}^{-1}$	
Celsius temperature	degree Celsius	°C	$T(°\text{C}) = T(\text{K}) - 273.15$	
Luminous flux	lumen	lm	$1\,\text{lm} = 1\,\text{cd}\,\text{sr}$	(1)
Illuminance	lux	lx	$1\,\text{lx} = 1\,\text{lm}\,\text{m}^{-2}$	(1)
Activity (of radio-nuclides)	becquerel	Bq	$1\,\text{Bq} = 1\,\text{s}^{-1}$	(2)
Absorbed dose	gray	Gy	$1\,\text{Gy} = 1\,\text{J}\,\text{kg}^{-1}$	(2)
Dose equivalent	sievert	Sv	$1\,\text{Sv} = 1\,\text{J}\,\text{kg}^{-1}$	(2)

1. *Luminous flux* and *illuminance* are derived quantities of photometry. The fundamental quantity and unit of photometry are the *luminous intensity* and the *candela*, respectively. The luminous flux is the flux of radiated energy, weighted by the average sensitivity curve of the human eye. The illuminance is the luminous flux incident on the unit surface of a body.
2. *Activity*, *absorbed dose*, and *dose equivalent* are quantities of dosimetry. Dosimetry deals with measuring intensity and effects of ionizing radiations. The activity is the number of radioactive decays per unit time. The absorbed dose is the energy released by the ionizing radiation to the unit mass of traversed substance. The dose equivalent takes into account the biological effectiveness of different kinds of ionizing radiation, for the same absorbed dose.

Non-SI Units Accepted for Use

In 1996, some non-SI units, largely used in science, technology, and common life, have been listed as *accepted for use* by the CIPM *(Comité International des Poids et Mesures)*. These units are divided into three categories and are listed in the following tables B.2, B.3, and B.4.

Table B.2. Units of frequent use.

Quantity	Unit	Symbol	Conversion
Volume	liter	l, L	$1\,\text{L} = 10^{-3}\,\text{m}^3$
Mass	metric ton	t	$1\,\text{t} = 10^3\,\text{kg}$
Time	minute	min	$1\,\text{min} = 60\,\text{s}$
Time	hour	h	$1\,\text{h} = 3600\,\text{s}$
Time	day	d	$1\,\text{d} = 86400\,\text{s}$
Plane angle	degree	°	$1° = (\pi/180)\,\text{rad}$
Plane angle	minute	′	$1' = (\pi/10800)\,\text{rad}$
Plane angle	second	″	$1'' = (\pi/648000)\,\text{rad}$
	neper	Np	$1\,\text{Np} = 1$
	bell	Bp	$1\,\text{B} = (1/2)\ln 10\,(\text{Np})$

Table B.3. Units whose SI value is obtained experimentally.

Quantity	Unit	Symbol	Approximate Conversion
Length	astronomical unit	au	$1\,\text{au} = 1.496 \times 10^{11}\,\text{m}$
Mass	atomic mass unit	u	$1\,\text{u} = 1.66 \times 10^{-27}\,\text{kg}$
Energy	electronvolt	eV	$1\,\text{eV} = 1.602 \times 10^{-19}\,\text{J}$

Table B.4. Units accepted in specific fields.

Quantity	Unit	Symbol	Conversion
Length	ångström	Å	$1\,\text{Å} = 10^{-10}\,\text{m}$
Length	nautical mile		$1852\,\text{m}$
Velocity	knot		$0.5144\,\text{m}\,\text{s}^{-1}$
Surface area	are	a	$1\,\text{a} = 10^2\,\text{m}^2$
Surface area	hectare	ha	$1\,\text{ha} = 10^4\,\text{m}^2$
Surface area	barn	b	$1\,\text{b} = 10^{-28}\,\text{m}^2$
Pressure	bar	bar	$1\,\text{bar} = 10^5\,\text{Pa}$

SI Prefixes

The SI codifies the names and the symbols of a series of prefixes for the decimal multiples and submultiples of units. The following Table B.5 lists the approved prefixes.

Table B.5. SI prefixes.

Factor	Name	Symbol	Factor	Name	Symbol
10^{24}	yotta-	Y-	10^{-24}	yocto-	y-
10^{21}	zetta-	Z-	10^{-21}	zepto-	z-
10^{18}	exa-	E-	10^{-18}	atto-	a-
10^{15}	peta-	P-	10^{-15}	femto-	f-
10^{12}	tera-	T-	10^{-12}	pico-	p-
10^{9}	giga-	G-	10^{-9}	nano-	n-
10^{6}	mega-	M-	10^{-6}	micro-	μ-
10^{3}	chilo-	k-	10^{-3}	milli-	m-
10^{2}	etto-	h-	10^{-2}	centi-	c-
10	deca-	da-	10^{-1}	deci-	d-

The prefix precedes the name of the base or derived unit. For example, $1\,\text{km} = 10^3\,\text{m}$; $1\,\mu\text{F} = 10^{-6}\,\text{F}$.

As an exception, multiples and submultiples of the unit of mass, the kilogram, are formed by attaching the prefix to the unit name "gram", and the prefix symbols to the symbol "g". For example, $1\,\text{mg} = 10^{-3}\,\text{g} = 10^{-6}\,\text{kg}$.

Rules for Writing SI Names and Symbols

1. The unit names are written in lower-case, without accents (for example, ampere, not Ampère).
2. The unit names have no plural (3 ampere, not 3 amperes).
3. In general, unit symbols are written in lower-case, but if the unit name is derived from the proper name of a person, the first letter of the symbol is a capital (mol for the mole, K for the kelvin).
4. Unit symbols are not followed by a period (except as normal punctuation at the end of a sentence).
5. Symbols always follow numerical values (1 kg, not kg 1).
6. The product of two or more units is expressed by a half-high dot or space (N·m or N m).
7. The ratio of two units is expressed by an oblique stroke or a negative exponent (J/s or J s^{-1}).

B.2 Units Not Accepted by the SI

In Table B.6, some units still often used for practical purposes, but not accepted by the SI, are listed for convenience.

Table B.6. Some non-SI units still frequently used.

Quantity	Unit	Symbol	Conversion
Linear density (textiles)	tex	tex	$10^{-6}\,\mathrm{kg\,m^{-1}}$
Mass	metric carat		$2\times10^{-4}\,\mathrm{kg}$
Force	force-kilogram	kgf	$9.80665\,\mathrm{N}$
Pressure	torr	torr	$133.322\,\mathrm{Pa}$
	atmosphere	atm	$101325\,\mathrm{Pa}$
Blood pressure	mercury millimeter	mm Hg	$133.322\,\mathrm{Pa}$
Energy	internat. calorie	cal	$4.1855\,\mathrm{J}$
Luminance	stilb	sb	$10^4\,\mathrm{nt}$
Kinematic viscosity	stokes	St	$10^{-4}\,\mathrm{m^2\,s^{-1}}$
Dynamic viscosity	poise	P	$10^{-1}\,\mathrm{Pa\,s}$
Activity	curie	Ci	$3.7\times10^{10}\,\mathrm{Bq}$
Absorbed dose	rad	rd	$10^{-2}\,\mathrm{Gy}$
Dose equivalent	rem	rem	$10^{-2}\,\mathrm{Sv}$
Exposure	roentgen	R	$2.58\times10^{-4}\,\mathrm{C\,kg^{-1}}$

B.3 British Units

Some of the more common British units are listed in the following Table B.7. Some units, in spite of the same name, have different values in the United Kingdom (UK) and in the United States of America (USA).

Table B.7. Some British units and their conversion to SI.

Quantity	Unit	Symbol	Conversion
Length	inch	in	25.4 mm
	foot	ft	304.8 mm
	yard	yd	0.9144 m
	statute mile	mi	1609.344 m
	nautical mile	naut mi	1853.184 m
Volume	cubic inch	in^3	16.387 cm^3
	fluid ounce UK	fl oz UK	28.413 cm^3
	fluid ounce USA	fl oz USA	29.574 cm^3
	pint UK	pt	568.261 cm^3
	liquid pint USA	liq pt	473.176 cm^3
	gallon UK	gal UK	4.5461 dm^3
	gallon USA	gal USA	3.7854 dm^3
	oil barrel		158.987 dm^3
Mass	ounce	oz	28.349 g
	pound	lb	0.4536 kg
Force	pound-force	lbf	4.448 N
Pressure	pound-force/square-inch	psi	6894.76 Pa
Energy	pound-force foot	lbf ft	1.3557 J
	British thermal unit	Btu	1054.5 J
	therm	therm	105.506 MJ
Power	horse power	hp	745.7 W
Temperature	degree Fahrenheit	°F	(5/9) K

B.4 Non-SI Units Currently Used in Physics

Unit	Symbol	Quantity	Approximate Conversion	Notes
angström	Å	length (atom. phys.)	10^{-10} m	
fermi	fm	length (nucl. phys.)	10^{-15} m	
astronomical unit	au	length (astron.)	1.496×10^{11} m	
light year		length (astron.)	9.46×10^{15} m	(1)
parsec	pc	length (astron.)	3.086×10^{16} m	(2)
barn	b	cross section	10^{-28} m^2	
inverse centimeters	cm^{-1}	wave-number	$100\,\text{m}^{-1}$	(3)
atomic mass unit	u	mass	1.66×10^{-27} Kg	
hartree	Hartree	energy	27.2 eV 4.36×10^{-18} J	(4)
rydberg	Ry	energy	13.6 eV 2.18×10^{-18} J	(4)
mercury centimeters	mm Hg	pressure	133.322 Pa	
röntgen	R	exposure	$2.58 \times 10^{-4}\,\text{C}\,\text{kg}^{-1}$	

1. The *light year* is the distance covered by electromagnetic radiation in vacuum in one tropic year (say in the time interval between two consecutive passages, in the same direction, of the Sun through the terrestrial equatorial plane).
2. The *parsec* (parallax second) corresponds to the distance at which the mean radius of the Earth orbit subtends an angle of 1″ (1″ = 4.84814×10^{-6} rad).
3. The *wave-number* is the inverse of the wavelength λ. The wave-number is connected to the frequency ν by the relation $\nu = v(1/\lambda)$, where v is the velocity of the wave.
4. *Hartree* and *rydberg* are natural units of energy, defined with reference to the ground state of the hydrogen atom. One Hartree corresponds to the absolute value of the potential energy of the electron in its ground state, say in SI units, $U_0 = -(1/4\pi\epsilon_0)(e^2/a_0)$ where a_0 is the radius of the first orbit in the Bohr model. 1 Ry = 0.5 Hartree corresponds to the ionization energy of the hydrogen atom.

B.5 Gauss cgs Units

Various cgs systems have been introduced, according to the law used to define the electromagnetic units as a function of mechanical units. The Gauss symmetrized cgs system uses the electrostatic cgs units for electrical quantities and the electromagnetic cgs units for magnetic quantities.

Table B.8. Some units of the cgs Gauss system.

Quantity	Unit	Symbol	Conversion
Force	dyne	dyn	$1\,\text{dyn} = 10^{-5}\,\text{N}$
Work, energy	erg	erg	$1\,\text{erg} = 10^{-7}\,\text{J}$
Electric charge	statcoulomb	statC	$1\,\text{statC} = 3.\bar{3}\times 10^{-10}\,\text{C}$
Electric current	statampere	statA	$1\,\text{statA} = 3.\bar{3}\times 10^{-10}\,\text{A}$
Electric potential	statvolt	statV	$1\,\text{statV} = 300\,\text{V}$
Magnetic flux density	gauss	G	$1\,\text{G} = 10^{-4}\,\text{T}$
Magnetic field	oersted	Oe	$1\,\text{Oe} = (1/4\pi)\times 10^{3}\,\text{A m}^{-1}$

C Tables

C.1 Greek Alphabet

Table C.1. The Greek alphabet.

Name	Lower-Case	Upper-Case	Name	Lower-Case	Upper-Case
Alpha	α	A	Nu	ν	N
Beta	β	B	Xi	ξ	Ξ
Gamma	γ	Γ	Omicron	o	O
Delta	δ	Δ	Pi	π	Π
Epsilon	ϵ, ε	E	Rho	ρ	P
Zeta	ζ	Z	Sigma	σ	Σ
Eta	η	H	Tau	τ	T
Theta	θ, ϑ	Θ	Upsilon	υ	Υ
Iota	ι	I	Phi	ϕ	Φ
Kappa	κ	K	Chi	χ	X
Lambda	λ	Λ	Psi	ψ	Ψ
Mu	μ	M	Omega	ω	Ω

C.2 Some Fundamental Constants of Physics

Some constant quantities are particularly important in physics, and are referred to as *fundamental constants*. Examples of fundamental constants are the velocity of light in vacuum, the electron mass, and the Avogadro number. The fundamental constants are measured in different laboratories and with different techniques, and the accuracy of their values progressively increases with time. An international committee, the CODATA (Committee on Data for Science and Technology), founded in 1966, gathers and critically compares the results obtained by the various laboratories. Periodically, CODATA issues a list of recommended values of the fundamental constants. The first list was issued in 1973, the second one in 1986, and the third one in 1998.

The values of some fundamental constants, taken from the 1998 compilation of CODATA, are listed in Table C.2. The values of the first three constants are exact. For the other constants, the absolute uncertainties are expressed in a form particularly suitable for very accurate measurements. The significant digits representing the uncertainty δX (typically 2) are written in parentheses immediately after the central value X_0; it is understood that the uncertainty refers to the corresponding least significant digits of the central value X_0. For example, for the electron mass,

$$m_e = 9.109\,381\,88(72) \times 10^{-31} \text{ kg}$$

stands for

$$m_e = (9.109\,381\,88 \pm 0.000\,000\,54) \times 10^{-31} \text{ kg}.$$

The complete list of values of fundamental constants, together with a description of the criteria used by CODATA for its compilation, can be found in the paper: *CODATA Recommended Values of the Fundamental Physical Constants: 1998*, by P. J. Mohr and B. N. Taylor, in *Reviews of Modern Physics*, vol. 72, pages 351–495 (2000).

Table C.2. Values of some fundamental constants of physics, from the 1998 compilation of CODATA.

Constant	Symbol	Value	Unit
Speed of light in vacuum	c	$299\,792\,458$	m s^{-1}
Vacuum permeability	μ_0	$4\pi \cdot 10^{-7}$	H m^{-1}
Vacuum permittivity	$\epsilon_0 = 1/\mu_0 c^2$	$8.854\,187\,817\ldots \cdot 10^{-12}$	F m^{-1}
Gravitational constant	G	$6.673(10) \cdot 10^{-11}$	m^{3}kg^{-1}s^{-2}
Planck constant	h	$6.626\,068\,76(52) \cdot 10^{-34}$	J s
Elementary charge	e	$1.602\,176\,462(63) \cdot 10^{-19}$	C
Electron mass	m_e	$9.109\,381\,88(72) \cdot 10^{-31}$	kg
Proton mass	m_p	$1.672\,621\,58(13) \cdot 10^{-27}$	kg
Neutron mass	m_n	$1.674\,927\,16(13) \cdot 10^{-27}$	kg
Atomic mass unit	u	$1.660\,538\,73(13) \cdot 10^{-27}$	kg
Fine structure constant	α	$7.297\,352\,533(27) \cdot 10^{-3}$	
Rydberg constant	R_∞	$109\,737\,31.568\,549(83)$	m^{-1}
Bohr radius	a_0	$0.529\,177\,2083(19) \cdot 10^{-10}$	m
Bohr magneton	μ_B	$927.400\,899(37) \cdot 10^{-26}$	J T^{-1}
Nuclear magneton	μ_N	$5.050\,783\,17(20) \cdot 10^{-27}$	J T^{-1}
Avogadro number	N_A	$6.022\,141\,99(47) \cdot 10^{23}$	mol^{-1}
Faraday constant	F	$96\,485.3415(39)$	C mol^{-1}
Molar gas constant	R	$8.314\,472(15)$	J mol^{-1} K^{-1}
Boltzmann constant	k_B	$1.380\,6503(24) \cdot 10^{-23}$	J K^{-1}

C.3 Integrals of the Standard Normal Distribution

The probability that a random variable x, normally distributed with mean m and standard deviation σ, is included between $x = \alpha$ and $x = \beta$ (Sect. 6.5) is:

$$P(\alpha < x < \beta) = \frac{1}{\sigma\sqrt{2\pi}} \int_\alpha^\beta \exp\left[-\frac{(x-m)^2}{2\sigma^2}\right] dx . \tag{C.1}$$

The calculation is simplified by defining the *standard normal variable*

$$z = \frac{x-m}{\sigma}, \tag{C.2}$$

which represents the deviation of x with respect to the mean m, measured in units σ. The distribution of z,

$$\phi(z) = \frac{1}{\sqrt{2\pi}} \exp\left[-\frac{z^2}{2}\right], \tag{C.3}$$

is the *standard normal density* (Scct. 6.5).

Once the limits of the integral have been substituted,

$$\alpha \to z_\alpha = \frac{\alpha - m}{\sigma}, \qquad \beta \to z_\beta = \frac{\beta - m}{\sigma}, \tag{C.4}$$

the calculation of probability reduces to

$$P(\alpha < x < \beta) = P(z_\alpha < z < z_\beta) = \frac{1}{\sqrt{2\pi}} \int_{z_\alpha}^{z_\beta} \exp\left[-\frac{z^2}{2}\right] dz . \tag{C.5}$$

To evaluate the integral in C.5, one uses tabulated values. Table C.3 gives the values

$$\Phi(z) = \frac{1}{\sqrt{2\pi}} \int_{-\infty}^z \exp\left[-\frac{z'^2}{2}\right] dz' , \tag{C.6}$$

and Table C.4 gives the values

$$\Phi^*(z) = \frac{1}{\sqrt{2\pi}} \int_0^z \exp\left[-\frac{z'^2}{2}\right] dz' . \tag{C.7}$$

In both tables, the first column gives the first two significant digits of z, and the first row gives the third significant digit. The main body of the tables gives the corresponding probability values.

Table C.3. Values of the integral $\Phi(z) = \int_{-\infty}^{z} \phi(z')\,dz'$.

z	0.00	0.01	0.02	0.03	0.04	0.05	0.06	0.07	0.08	0.09
-3.8	0.0001									
-3.6	0.0002									
-3.4	0.0003									
-3.2	0.0007									
-3.0	0.0014									
-2.9	0.0019									
-2.8	0.0026									
-2.7	0.0035									
-2.6	0.0047									
-2.5	0.0062									
-2.4	0.0082									
-2.3	0.0107									
-2.2	0.0139									
-2.1	0.0179									
-2.0	0.0228									
-1.9	0.0288	0.0281	0.0274	0.0268	0.0262	0.0256	0.0250	0.0244	0.0239	0.0233
-1.8	0.0359	0.0351	0.0344	0.0336	0.0329	0.0322	0.0314	0.0307	0.0301	0.0294
-1.7	0.0446	0.0436	0.0427	0.0418	0.0409	0.0401	0.0392	0.0384	0.0375	0.0367
-1.6	0.0548	0.0537	0.0526	0.0516	0.0505	0.0495	0.0485	0.0475	0.0465	0.0455
-1.5	0.0668	0.0655	0.0643	0.0630	0.0618	0.0606	0.0594	0.0582	0.0571	0.0559
-1.4	0.0808	0.0793	0.0778	0.0764	0.0749	0.0735	0.0721	0.0708	0.0694	0.0681
-1.3	0.0968	0.0951	0.0934	0.0918	0.0901	0.0885	0.0869	0.0853	0.0838	0.0823
-1.2	0.1151	0.1131	0.1112	0.1093	0.1075	0.1056	0.1038	0.1020	0.1003	0.0985
-1.1	0.1357	0.1335	0.1314	0.1292	0.1271	0.1251	0.1230	0.1210	0.1190	0.1170
-1.0	0.1587	0.1563	0.1539	0.1515	0.1492	0.1469	0.1446	0.1423	0.1401	0.1379
-0.9	0.1841	0.1814	0.1788	0.1762	0.1736	0.1711	0.1685	0.1660	0.1635	0.1611
-0.8	0.2119	0.2090	0.2061	0.2033	0.2005	0.1977	0.1949	0.1922	0.1894	0.1867
-0.7	0.2420	0.2389	0.2358	0.2327	0.2297	0.2266	0.2236	0.2206	0.2177	0.2148
-0.6	0.2743	0.2709	0.2676	0.2643	0.2611	0.2578	0.2546	0.2514	0.2483	0.2451
-0.5	0.3085	0.3050	0.3015	0.2981	0.2946	0.2912	0.2877	0.2843	0.2810	0.2776
-0.4	0.3646	0.3409	0.3372	0.3336	0.3300	0.3264	0.3228	0.3192	0.3156	0.3121
-0.3	0.3821	0.3783	0.3745	0.3707	0.3669	0.3632	0.3594	0.3557	0.3520	0.3483
-0.2	0.4207	0.4168	0.4129	0.4090	0.4052	0.4013	0.3974	0.3936	0.3897	0.3859
-0.1	0.4602	0.4562	0.4522	0.4483	0.4443	0.4404	0.4364	0.4325	0.4286	0.4247
-0.0	0.5000	0.4960	0.4920	0.4880	0.4840	0.4801	0.4761	0.4721	0.4681	0.4641

(continued)

C.3 Integrals of the Standard Normal Distribution

Table C.3. (continued)

z	0.00	0.01	0.02	0.03	0.04	0.05	0.06	0.07	0.08	0.09
+0.0	0.5000	0.5040	0.5080	0.5120	0.5160	0.5199	0.5239	0.5279	0.5319	0.5359
+0.1	0.5398	0.5438	0.5478	0.5517	0.5557	0.5596	0.5636	0.5675	0.5714	0.5753
+0.2	0.5793	0.5832	0.5871	0.5910	0.5948	0.5987	0.6026	0.6064	0.6103	0.6141
+0.3	0.6179	0.6217	0.6255	0.6293	0.6331	0.6368	0.6406	0.6443	0.6480	0.6517
+0.4	0.6554	0.6591	0.6628	0.6664	0.6700	0.6736	0.6772	0.6808	0.6844	0.6879
+0.5	0.6915	0.6950	0.6985	0.7019	0.7054	0.7088	0.7123	0.7157	0.7190	0.7224
+0.6	0.7257	0.7291	0.7324	0.7357	0.7389	0.7422	0.7454	0.7486	0.7517	0.7549
+0.7	0.7580	0.7611	0.7642	0.7673	0.7703	0.7734	0.7764	0.7794	0.7823	0.7852
+0.8	0.7881	0.7910	0.7939	0.7967	0.7995	0.8023	0.8051	0.8078	0.8106	0.8133
+0.9	0.8159	0.8186	0.8212	0.8238	0.8264	0.8289	0.8315	0.8340	0.8365	0.8389
+1.0	0.8413	0.8437	0.8461	0.8485	0.8508	0.8531	0.8554	0.8577	0.8599	0.8621
+1.1	0.8643	0.8665	0.8686	0.8708	0.8729	0.8749	0.8770	0.8790	0.8810	0.8830
+1.2	0.8849	0.8869	0.8888	0.8907	0.8925	0.8944	0.8962	0.8980	0.8997	0.9015
+1.3	0.9032	0.9049	0.9066	0.9082	0.9099	0.9115	0.9131	0.9147	0.9162	0.9177
+1.4	0.9192	0.9207	0.9222	0.9236	0.9251	0.9265	0.9279	0.9292	0.9306	0.9319
+1.5	0.9332	0.9345	0.9357	0.9370	0.9382	0.9394	0.9406	0.9418	0.9429	0.9441
+1.6	0.9452	0.9463	0.9474	0.9484	0.9495	0.9505	0.9515	0.9525	0.9535	0.9545
+1.7	0.9554	0.9564	0.9573	0.9582	0.9591	0.9599	0.9608	0.9616	0.9625	0.9633
+1.8	0.9641	0.9649	0.9656	0.9664	0.9671	0.9678	0.9686	0.993	0.9699	0.9706
+1.9	0.9713	0.9719	0.9726	0.9732	0.9738	0.9744	0.9750	0.9756	0.9761	0.9767
+2.0	0.9772									
+2.1	0.9821									
+2.2	0.9861									
+2.3	0.9893									
+2.4	0.9918									
+2.5	0.9938									
+2.6	0.9953									
+2.7	0.9965									
+2.8	0.9974									
+2.9	0.9981									
+3.0	0.9986									
+3.2	0.9993									
+3.4	0.9997									
+3.6	0.9998									
+3.8	0.9999									

Table C.4. Values of the integral $\Phi^*(z) = \int_0^z \phi(z')\,dz'$.

z	0.00	0.01	0.02	0.03	0.04	0.05	0.06	0.07	0.08	0.09
0.0	0.0000	0.0040	0.0080	0.0120	0.0160	0.0199	0.0239	0.279	0.0319	0.0359
0.1	0.0398	0.0438	0.0478	0.0517	0.0557	0.0596	0.0636	0.0675	0.0714	0.0753
0.2	0.0793	0.0832	0.0871	0.0910	0.0948	0.0987	0.1026	0.1064	0.1103	0.1141
0.3	0.1179	0.1217	0.1255	0.1293	0.1331	0.1368	0.1406	0.1443	0.1480	0.1517
0.4	0.1554	0.1591	0.1628	0.1664	0.1700	0.1736	0.1772	0.1808	0.1844	0.1879
0.5	0.1915	0.1950	0.1985	0.2019	0.2054	0.2088	0.2123	0.2157	0.2190	0.2224
0.6	0.2257	0.2291	0.2324	0.2357	0.2389	0.2422	0.2454	0.2486	0.2517	0.2549
0.7	0.2580	0.2611	0.2642	0.2673	0.2704	0.2734	0.2764	0.2794	0.2823	0.2852
0.8	0.2881	0.2910	0.2939	0.2967	0.2995	0.3023	0.3051	0.3078	0.3106	0.3133
0.9	0.3159	0.3186	0.3212	0.3238	0.3264	0.3289	0.3315	0.3340	0.3365	0.3389
1.0	0.3413	0.3438	0.3461	0.3485	0.3508	0.3531	0.3554	0.3577	0.3599	0.3621
1.1	0.3643	0.3665	0.3686	0.3708	0.3729	0.3749	0.3770	0.3790	0.3810	0.3830
1.2	0.3849	0.3869	0.3888	0.3907	0.3925	0.3944	0.3962	0.3980	0.3997	0.9015
1.3	0.4032	0.4049	0.4066	0.4082	0.4099	0.4115	0.4131	0.4147	0.4162	0.4177
1.4	0.4192	0.4207	0.4222	0.4236	0.4251	0.4265	0.4279	0.4292	0.4306	0.4319
1.5	0.4332	0.4345	0.4357	0.4370	0.4382	0.4394	0.4406	0.4418	0.4429	0.4441
1.6	0.4452	0.4463	0.4474	0.4484	0.4495	0.4505	0.4515	0.4525	0.4535	0.4545
1.7	0.4554	0.4564	0.4573	0.4582	0.4591	0.4599	0.4608	0.4616	0.4625	0.4633
1.8	0.4641	0.4649	0.4656	0.4664	0.4671	0.4678	0.4686	0.4693	0.4699	0.4706
1.9	0.4713	0.4719	0.4726	0.4732	0.4738	0.4744	0.4750	0.4756	0.4761	0.4767
2.0	0.4772	0.4778	0.4783	0.4788	0.4793	0.4798	0.4803	0.4808	0.4812	0.4817
2.1	0.4821	0.4826	0.4830	0.4834	0.4838	0.4842	0.4846	0.4850	0.4854	0.4857
2.2	0.4861	0.4864	0.4868	0.4871	0.4875	0.4878	0.4881	0.4884	0.4887	0.4890
2.3	0.4893	0.4896	0.4898	0.4901	0.4904	0.4906	0.4909	0.4911	0.4913	0.4916
2.4	0.4918	0.4920	0.4922	0.4925	0.4927	0.4929	0.4931	0.4932	0.4934	0.4936
2.5	0.4938	0.4940	0.4941	0.4943	0.4945	0.4946	0.4948	0.4949	0.4951	0.4952
2.6	0.4953	0.4955	0.4956	0.4957	0.4959	0.4960	0.4961	0.4962	0.4963	0.4964
2.7	0.4965	0.4966	0.4967	0.4968	0.4969	0.4970	0.4971	0.4972	0.4973	0.4974
2.8	0.4974	0.4975	0.4776	0.4977	0.4977	0.4978	0.4979	0.4979	0.4980	0.4981
2.9	0.4981	0.4982	0.4982	0.4983	0.4984	0.4984	0.4985	0.4985	0.4986	0.4986
3.0	0.4987									
3.5	0.4998									
4.0	0.4999									

C.4 Integrals of the Student Distribution

If a random variable X is sampled N times, the probability $\mathcal{P}'$ that an interval of half-width $k\tilde{\sigma}[m^*]$, centered on the sample mean m^*, includes the parent mean m,

$$\mathcal{P}' = \mathcal{P}\{|m - m^*| < k\tilde{\sigma}[m^*]\}, \tag{C.8}$$

is called the confidence level relative to the coverage factor k (Sect. 9.1). If the variable X is normally distributed, the confidence level corresponding to a given coverage factor k can be calculated by integrating the Student distribution $S_\nu(t)$ (Sect. 9.2), where $t = (m^* - m)/\tilde{\sigma}[m^*]$, and $\nu = N - 1$ is the number of degrees of freedom:

$$\mathcal{P}'_k = 2 \int_0^k S_\nu(t)\, dt. \tag{C.9}$$

The values of the integral as a function of k and ν are listed in Table C.5, and the values of k as a function of $\mathcal{P}'$ and ν are listed in Table C.6.

Table C.5. The table gives the percent confidence levels $\mathcal{P}'$ for selected values of coverage factor k (first row) and degrees of freedom ν (first column). The value $k = 1$ corresponds to the standard uncertainty, measured by the estimate of the standard deviation of the sample means $\delta X = \tilde{\sigma}[m^*]$. The value $\nu = \infty$ corresponds to the asymptotic limit, where the Student distribution $S(t)$ becomes equal to the standard normal distribution $\phi(z)$.

$k \rightarrow$	1	1.5	2	2.5	3
$\nu = 1$	50.00	62.57	70.48	75.78	79.52
$\nu = 2$	57.74	72.76	81.65	87.04	90.45
$\nu = 3$	60.90	76.94	86.07	91.23	94.23
$\nu = 4$	62.61	79.20	88.39	93.32	96.01
$\nu = 5$	63.68	80.61	89.81	94.55	96.99
$\nu = 6$	64.41	81.57	90.76	95.35	97.60
$\nu = 7$	64.94	82.27	91.44	95.90	98.01
$\nu = 8$	65.34	82.80	91.95	96.31	98.29
$\nu = 9$	65.66	83.21	92.34	96.61	98.50
$\nu = 10$	65.91	83.55	92.66	96.86	98.67
$\nu = 15$	66.68	84.56	93.61	97.55	99.10
$\nu = 20$	67.07	85.08	94.07	97.88	99.29
$\nu = 30$	67.47	85.59	94.54	98.19	99.46
$\nu = 40$	67.67	85.85	94.77	98.34	99.54
$\nu = 50$	67.79	86.01	94.91	98.43	99.58
$\nu = 100$	68.03	86.32	95.18	98.60	99.66
$\nu = \infty$	68.27	86.64	95.45	98.76	99.73

Table C.6. The table gives the values of the coverage factor k for selected values of percent confidence level $\mathcal{P}'$ (first row) and degrees of freedom ν (first column). The value $\nu = \infty$ corresponds to the asymptotic limit, where the Student distribution $S(t)$ becomes equal to the standard normal distribution $\phi(z)$. The confidence levels $\mathcal{P}' = 68.27\%$, 95.45%, and 99.73% correspond, for $\nu = \infty$, to coverage factors $k = 1$, 2, and 3, respectively.

$\mathcal{P}'$ (%) →	50	68.27	90	95	95.45	99	99.73
$\nu=1$	1.000	1.84	6.31	12.71	13.97	63.66	235.80
$\nu=2$	0.816	1.32	2.92	4.30	4.53	9.92	19.21
$\nu=3$	0.765	1.20	2.35	3.18	3.31	5.84	9.22
$\nu=4$	0.741	1.14	2.13	2.78	2.87	4.60	6.62
$\nu=5$	0.727	1.11	2.02	2.57	2.65	4.03	5.51
$\nu=6$	0.718	1.09	1.94	2.45	2.52	3.71	4.90
$\nu=7$	0.711	1.08	1.89	2.36	2.43	3.50	4.53
$\nu=8$	0.706	1.07	1.86	2.31	2.37	3.36	4.28
$\nu=9$	0.703	1.06	1.83	2.26	2.32	3.25	4.09
$\nu=10$	0.700	1.05	1.81	2.23	2.28	3.17	3.96
$\nu=15$	0.691	1.03	1.75	2.13	2.18	2.95	3.59
$\nu=20$	0.687	1.03	1.72	2.09	2.13	2.85	3.42
$\nu=30$	0.683	1.02	1.70	2.04	2.09	2.75	3.27
$\nu=40$	0.681	1.01	1.68	2.02	2.06	2.70	3.20
$\nu=50$	0.680	1.01	1.68	2.01	2.05	2.68	3.16
$\nu=100$	0.678	1.005	1.660	1.98	2.02	2.63	3.08
$\nu=\infty$	0.674	1.000	1.645	1.96	2.00	2.58	3.00

C.5 Integrals of the Chi Square Distribution

The continuous random variable χ^2 is defined as a sum of squared standard normal random variables Z (Sect. 11.3):

$$\chi^2 = \sum_{k=1}^{\nu} Z^2 \,. \tag{C.10}$$

The parameter ν is the number of degrees of freedom.

The *reduced* $\tilde{\chi}^2$ is defined as

$$\tilde{\chi}^2 = \frac{\chi^2}{\nu} = \frac{1}{\nu}\sum_{k=1}^{\nu} Z^2 \,. \tag{C.11}$$

For every integer value of ν, one can calculate the distributions of both χ^2 and $\tilde{\chi}^2$. The probability density $f_\nu(\chi^2)$ is given by

$$f_\nu(\chi^2) = \frac{1}{2^{\nu/2}\,\Gamma(\nu/2)} \left(\chi^2\right)^{\nu/2-1} \exp\left[-\chi^2/2\right] \,, \tag{C.12}$$

where the Γ function is defined as follows.

For n integer $\quad \Gamma(n+1) = n!$
For n half-integer $\Gamma(n+1/2) = \sqrt{\pi}/2^n \,(2n-1)(2n-3)\cdots 5\times 3\times 1$.

The distributions of χ^2 and $\tilde{\chi}^2$, for different values of ν, are shown in Figs. 11.3 and 11.3 of Sect. 11.3.

The values of the integral of the reduced chi-squared,

$$\int_{\tilde{\chi}_0^2}^{\infty} f_\nu(\tilde{\chi}^2)\, \mathrm{d}\tilde{\chi}^2 = \mathcal{P}\left[\tilde{\chi}^2 > \tilde{\chi}_0^2\right] \,, \tag{C.13}$$

are listed in Table C.7. To facilitate reading, the values are listed as percent probabilities.

The meaning of the integral (C.13) and of Table C.7 is clarified by Fig. C.5, which refers to the case of $\nu = 5$ and $\tilde{\chi}_0^2 = 0.8$. The integral (C.13) measures the area under the curve of the probability density for $\tilde{\chi}^2 \geq \tilde{\chi}_0^2$ (shaded area in Fig. C.5, left), corresponding to $1 - F(\tilde{\chi}_0^2)$, where $F(\tilde{\chi}_0^2)$ is the cumulative distribution function for $\tilde{\chi}^2 = \tilde{\chi}_0^2$ (Fig. C.5, right).

Table C.7. Percent probability that the reduced chi square $\tilde{\chi}^2$ is higher than an experimentally observed value $\tilde{\chi}_0^2$. For selected degrees of freedom ν (first column) and reduced chi-squared $\tilde{\chi}_0^2$ (first row), the tables give the values of percent probabilities

$$100 \int_{\tilde{\chi}_0^2}^{\infty} f_\nu\left(\tilde{\chi}^2\right) d\tilde{\chi}^2 = 100\,\mathcal{P}\left[\tilde{\chi}^2 > \tilde{\chi}_0^2\right].$$

$\tilde{\chi}_0^2 \rightarrow$	0	0.5	1.0	1.5	2.0	2.5	3.0	3.5	4.0	4.5	5.0	5.5	6.0	8.0	10.0
$\nu=1$	100	48	32	22	16	11	8.3	6.1	4.6	3.4	2.5	1.9	1.4	0.5	0.2
$\nu=2$	100	61	37	22	14	8.2	5.0	3.0	1.8	1.1	0.7	0.4	0.2		
$\nu=3$	100	68	39	21	11	5.8	2.9	1.5	0.7	0.4	0.2	0.1			
$\nu=4$	100	74	41	20	9.2	4.0	1.7	0.7	0.3	0.1	0.1				
$\nu=5$	100	78	42	19	7.5	2.9	1.0	0.4	0.1						

$\tilde{\chi}_0^2 \rightarrow$	0	0.2	0.4	0.6	0.8	1.0	1.2	1.4	1.6	1.8	2.0	2.2	2.4	2.6	2.8	3.0
$\nu=1$	100	65	53	44	37	32	27	24	21	18	16	14	12	11	9.4	8.3
$\nu=2$	100	82	67	55	45	37	30	25	20	17	14	11	9.1	7.4	6.1	5.0
$\nu=3$	100	90	75	61	49	39	31	24	19	14	11	8.6	6.6	5.0	3.8	2.9
$\nu=4$	100	94	81	66	52	41	31	23	17	13	9.2	6.6	4.8	3.4	2.4	1.7
$\nu=5$	100	96	85	70	55	42	31	22	16	11	7.5	5.1	3.5	2.3	1.6	1.0
$\nu=6$	100	98	88	73	57	42	30	21	14	9.5	6.2	4.0	2.5	1.6	1.0	0.6
$\nu=7$	100	99	90	76	59	43	30	20	13	8.2	5.1	3.1	1.9	1.1	0.7	0.4
$\nu=8$	100	99	92	78	60	43	29	19	12	7.2	4.2	2.4	1.4	0.8	0.4	0.2
$\nu=9$	100	99	94	80	62	44	29	18	11	6.3	3.5	1.9	1.0	0.5	0.3	0.1
$\nu=10$	100	100	95	82	63	44	29	17	10	5.5	2.9	1.5	0.8	0.4	0.2	0.1
$\nu=11$	100	100	96	83	64	44	28	16	9.1	4.8	2.4	1.2	0.6	0.3	0.1	0.1
$\nu=12$	100	100	96	84	65	45	28	16	8.4	4.2	2.0	0.9	0.4	0.2	0.1	
$\nu=13$	100	100	97	86	66	45	27	15	7.7	3.7	1.7	0.7	0.3	0.1	0.1	
$\nu=14$	100	100	98	87	67	45	27	14	7.1	3.3	1.4	0.6	0.2	0.1		
$\nu=15$	100	100	98	88	68	45	26	14	6.5	2.9	1.2	0.5	0.2	0.1		
$\nu=16$	100	100	98	89	69	45	26	13	6.0	2.5	1.0	0.4	0.1			
$\nu=17$	100	100	99	90	70	45	25	12	5.5	2.2	0.8	0.3	0.1			
$\nu=18$	100	100	99	90	70	46	25	12	5.1	2.0	0.7	0.2	0.1			
$\nu=19$	100	100	99	91	71	46	25	11	4.7	1.7	0.6	0.2	0.1			
$\nu=20$	100	100	99	92	72	46	24	11	4.3	1.5	0.5	0.1				
$\nu=22$	100	100	99	93	73	46	23	10	3.7	1.2	0.4	0.1				
$\nu=24$	100	100	100	94	74	46	23	9.2	3.2	0.9	0.3	0.1				
$\nu=26$	100	100	100	95	75	46	22	8.5	2.7	0.7	0.2					
$\nu=28$	100	100	100	95	76	46	21	7.8	2.3	0.6	0.1					
$\nu=30$	100	100	100	96	77	47	21	7.2	2.0	0.5	0.1					

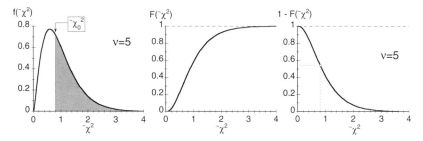

Fig. C.1. Distribution of the reduced chi square for $\nu = 5$. Left: probability density $f(\tilde{\chi}^2)$. Center: cumulative distribution function $F(\tilde{\chi}^2)$. Right: $1 - F(\tilde{\chi}^2)$.

C.6 Integrals of the Linear Correlation Coefficient Distribution

The *linear correlation coefficient* r of a number N of pairs (x_i, y_i) has been defined in Sect. 10.2 as

$$r = \frac{\sum_i (x_i - m_x^*)(y_i - m_y^*)}{\sqrt{\sum_i (x_i - m_x^*)^2}\sqrt{\sum_i (y_i - m_y^*)^2}}. \tag{C.14}$$

For two completely uncorrelated quantities, one can calculate the distribution of the random variable r. By integration, one can then calculate the probability

$$P_N\left(|r| \geq |r_o|\right), \tag{C.15}$$

where r_o is the experimental value.

The probability is calculated by the integral

$$P_N\left(|r| \geq |r_o|\right) = \frac{2\Gamma[(N-1)/2]}{\sqrt{\pi}\,\Gamma[(N-2)/2]} \int_{|r_o|}^{1} (1-r^2)^{(N-4)/4}\,dr, \tag{C.16}$$

where the Γ function is defined as follows.

For n integer $\quad \Gamma(n+1) = n!$
For n half-integer $\Gamma(n+1/2) = \sqrt{\pi}/2^n\,(2n-1)(2n-3)\cdots 5 \times 3 \times 1$.

Some values of the integral are listed in Table C.8.

Table C.8. The table gives the percent probability that, for N observations, the value r for two completely uncorrelated variables is larger in absolute value than the observed value r_o: $P_N(|r| \geq |r_o|)$.

$|r_o| \rightarrow$	0	0.1	0.2	0.3	0.4	0.5	0.6	0.7	0.8	0.9	1.0
$N=3$	100	94	87	81	74	67	59	51	41	29	0
$N=4$	100	90	80	70	60	50	40	30	20	10	0
$N=5$	100	87	75	62	50	39	28	19	10	3.7	0
$N=6$	100	85	70	56	43	31	21	12	5.6	1.4	0
$N=7$	100	83	67	51	37	25	15	8.0	3.1	0.6	0
$N=8$	100	81	63	47	33	21	12	5.3	1.7	0.2	0
$N=9$	100	80	61	43	29	17	8.8	3.6	1.0	0.1	0
$N=10$	100	78	58	40	25	14	6.7	2.4	0.5		0
$N=11$	100	77	56	37	22	12	5.1	1.6	0.3		0
$N=12$	100	76	53	34	20	9.8	3.9	1.1	0.2		0
$N=13$	100	75	51	32	18	8.2	3.0	0.8	0.1		0
$N=14$	100	73	49	30	16	6.9	2.3	0.5	0.1		0
$N=15$	100	72	47	28	14	5.8	1.8	0.4			0
$N=16$	100	71	46	26	12	4.9	1.4	0.3			0
$N=17$	100	70	44	24	11	4.1	1.1	0.2			0
$N=18$	100	69	43	23	10	3.5	0.8	0.1			0
$N=19$	100	68	41	21	9.0	2.9	0.7	0.1			0
$N=20$	100	67	40	20	8.1	2.5	0.5	0.1			0
$N=25$	100	63	34	15	4.8	1.1	0.2				0
$N=30$	100	60	29	11	2.9	0.5					0
$N=35$	100	57	25	8.0	1.7	0.2					0
$N=40$	100	54	22	6.0	1.1	0.1					0
$N=45$	100	51	19	4.5	0.6						0

D Mathematical Complements

D.1 Response of Instruments: Differential Equations

The dynamical behavior of many instruments (Sect. 3.5) can be approximately described by linear differential equations of the first-order (3.4),

$$a_1 \frac{dZ}{dt} + a_0 Z = b_0 X , \qquad (D.1)$$

or of the second-order (3.5),

$$a_2 \frac{d^2 Z}{dt^2} + a_1 \frac{dZ}{dt} + a_0 Z = b_0 X . \qquad (D.2)$$

On general grounds, the solution $Z(t)$ of a linear differential equation with constant coefficients is the sum of two functions:

$$Z(t) = Z_{tr}(t) + Z_{st}(t) , \qquad (D.3)$$

where

- $Z_{tr}(t)$ is the general solution of the homogeneous equation (say the equation where $X = 0$), and has *transient* character,
- $Z_{st}(t)$ is one particular solution of the nonhomogeneous equation, say the complete equation, and has *stationary* character.

The solution of the first- and second-order homogeneous equations, and the general solutions for the case of a step input function $X(t)$, are studied here.

Solution of the First-Order Homogeneous Equation

The first-order homogeneous equation

$$\frac{dZ}{dt} + \gamma Z = 0 , \quad (\gamma = a_0/a_1) \qquad (D.4)$$

can be solved by the separation of variables, $dZ/Z = -\gamma\, dt$, and the integration between $t = 0$, where $Z(t) = Z_0$, and the generic time t:

$$\int_{Z_0}^{Z} \frac{\mathrm{d}Z'}{Z'} = \int_{t'}^{t} \mathrm{d}t' . \tag{D.5}$$

The solution is
$$Z_{\mathrm{tr}}(t) = Z_0 \exp(-\gamma t) , \tag{D.6}$$
where the constant Z_0 depends on the initial conditions.

Solution of the Second-Order Homogeneous Equation

The second-order homogeneous differential equation is
$$\frac{\mathrm{d}^2 Z}{\mathrm{d}t^2} + 2\gamma \frac{\mathrm{d}Z}{\mathrm{d}t} + \omega_0^2 = 0 , \quad (2\gamma = a_1/a_2, \ \omega_0^2 = a_0/a_2) . \tag{D.7}$$

One expects a solution containing a mixed sinusoidal and exponential behavior (induced by the second-order and first-order terms, respectively). We thus try a complex solution
$$Z_{\mathrm{st}}(t) = \exp(\lambda t), \quad \text{where } \lambda = a + ib . \tag{D.8}$$

By substituting (D.8) into (D.7), one obtains the *characteristic equation*
$$\lambda^2 + 2\gamma \lambda + \omega_0^2 = 0 , \tag{D.9}$$
whose complex solutions are
$$\lambda_+ = -\gamma + \sqrt{\gamma^2 - \omega_0^2}, \quad \lambda_- = -\gamma - \sqrt{\gamma^2 - \omega_0^2} . \tag{D.10}$$

Three different cases can be distinguished: $\gamma < \omega_0$, $\gamma = \omega_0$, and $\gamma > \omega_0$.

First Case: $\gamma < \omega_0$

The radicand in (D.10) is negative. By introducing the new parameter
$$\omega_s = \sqrt{\omega_0^2 - \gamma^2} , \tag{D.11}$$
two linearly independent solutions of (D.7) are
$$Z_+(t) = e^{-\gamma t} e^{i\omega_s t} , \quad Z_-(t) = e^{-\gamma t} e^{-i\omega_s t} , \tag{D.12}$$
which can be easily transformed into two real solutions by linear combination:
$$z_1(t) = e^{-\gamma t} \cos \omega_s t , \quad z_2(t) = e^{-\gamma t} \sin \omega_s t . \tag{D.13}$$

The general solution of (D.7) for $\gamma < \omega_0$ is thus
$$Z_{\mathrm{tr}}(t) = a_1 z_1(t) + a_2 z_2(t) = Z_0 e^{-\gamma t} \sin(\omega_s t + \phi) , \tag{D.14}$$
where the parameters Z_0 and ϕ depend on the initial conditions.

D.1 Response of Instruments: Differential Equations

Second Case: $\gamma = \omega_0$ (Critical Damping)

The radicand in (D.10) is zero, so that (D.9) has two coincident solutions: $\lambda = -\gamma$. Two linearly independent solutions of (D.7) are

$$z_1(t) = e^{-\gamma t}, \quad z_2(t) = t\, e^{-\gamma t}, \tag{D.15}$$

and the general solution is

$$Z_{\text{tr}}(t) = Z_1 z_1(t) + Z_2 z_2(t) = (Z_1 + Z_2 t)\, e^{-\gamma t}, \tag{D.16}$$

where the parameters Z_1 and Z_2 depend on the initial conditions.

Third Case: $\gamma > \omega_0$

The radicand in (D.10) is positive. Let $\delta = \sqrt{\gamma^2 - \omega_0^2}$ ($\delta < \gamma$). Two linearly independent solutions of (D.7) are

$$z_1(t) = e^{-(\gamma-\delta)t}, \quad z_2(t) = e^{-(\gamma+\delta)t}, \tag{D.17}$$

and the general solution is

$$Z_{\text{tr}}(t) = Z_1 z_1(t) + Z_2 z_2(t) = Z_1\, e^{-(\gamma-\delta)t} + Z_2\, e^{-(\gamma+\delta)t}, \tag{D.18}$$

where the parameters Z_1 and Z_2 depend on the initial conditions.

Step Input for a First-Order Instrument

Let us now consider the complete equation (D.1) for a first-order instrument, and search for a solution in the particular case of a step input:

$$X(t) = \begin{cases} X_0 & \text{for } t < 0, \\ X_1 & \text{for } t > 0. \end{cases} \tag{D.19}$$

Let us suppose that, for $t < 0$, the instrument is stabilized at the stationary response $Z_{\text{st}}(t < 0) = (b_0/a_0)\, X_0$, and let us consider the behavior for $t > 0$. After a sufficiently long time interval, one expects that the transient solution becomes negligible, and the stationary solution is $Z_{\text{st}}(t \geq 0) = b_0/a_0\, X_1$. The general solution of (3.4) for $t \geq 0$ is thus

$$Z(t) = Z_{\text{tr}}(t) + Z_{\text{st}}(t) = Z_0\, e^{-t/\tau} + (b_0/a_0)\, X_1. \tag{D.20}$$

To determine the parameter Z_0, let us consider the time $t = 0$ and impose on (D.20) the initial condition $Z(0) = (b_0/a_0)\, X_0$:

$$Z(0) = (b_0/a_0)\, X_0 = Z_0 + (b_0/a_0)\, X_1, \tag{D.21}$$

whence $Z_0 = (b_0/a_0)\, (X_0 - X_1)$, so that (Fig. 3.10 of Sect. 3.5)

$$Z(t) = \frac{b_0}{a_0}(X_0 - X_1)\, e^{-t/\tau} + \frac{b_0}{a_0} X_1. \tag{D.22}$$

Step Input for a Second-Order Instrument

Let us now consider the complete equation (D.2) for a second-order instrument, and search again for a solution in the particular case of a step input (D.19). Also for the second-order instrument, the asymptotic stationary behavior will be $Z_{st}(t \geq 0) = b_0/a_0 \, X_1$. To determine the parameters of the transient solution, it is again necessary to consider three different cases (Fig. 3.11 of Sect. 3.5).

First Case: $\gamma < \omega_0$

The solution of (3.5) for the step input is

$$Z(t) = Z_0 \, e^{-\gamma t} \sin(\omega_s t + \phi) + \frac{b_0}{a_0} X_1 \,. \tag{D.23}$$

By imposing that, for $t = 0$, $Z = (b_0/a_0) X_0$ and $dZ/dt = 0$, one obtains:

$$Z_0 = \sqrt{(\gamma/\omega_s)^2 + 1} \, \frac{b_0}{a_0} (X_0 - X_1), \quad \phi = \text{arctg}(-\gamma/\omega_s) \,. \tag{D.24}$$

Second Case: $\gamma = \omega_0$ *(Critical Damping)*

The solution of (3.5) for the step input is

$$Z(t) = (Z_1 + Z_2 t) \, e^{-\gamma t} + \frac{b_0}{a_0} X_1 \,. \tag{D.25}$$

By imposing that, for $t = 0$, $Z = (b_0/a_0) X_0$ and $dZ/dt = 0$, one determines Z_1 and Z_2, and finally

$$Z(t) = \frac{b_0}{a_0} (X_0 - X_1) \, [1 - \gamma t] \, e^{-\gamma t} + \frac{b_0}{a_0} X_1 \,. \tag{D.26}$$

Third Case: $\gamma > \omega_0$

The solution of (3.5) for the step input is

$$Z(t) = Z_1 \, e^{-(\gamma - \delta)t} + Z_2 \, e^{-(\gamma + \delta)t} + \frac{b_0}{a_0} X_1 \,. \tag{D.27}$$

By imposing that, for $t = 0$, $Z = (b_0/a_0) X_0$ and $dZ/dt = 0$, one determines Z_1 and Z_2, and finally

$$Z_1 = \left[1 - \frac{\gamma - \delta}{\gamma + \delta}\right]^{-1} \left(\frac{b_0}{a_0}\right) (X_0 - X_1), \quad Z_2 = -\frac{\gamma - \delta}{\gamma + \delta} Z_1 \,. \tag{D.28}$$

D.2 Transformed Functions of Distributions

The study of the random variable distributions, including the calculation of their moments, is greatly facilitated by two functions, the moment generating function and the characteristic function.

Moment Generating Function

The moment generating function $\mathcal{G}(t)$ is a function of a real variable t.

(a) For a discrete random variable K, the moment generating function is:

$$\mathcal{G}_k(t) = \langle e^{tk} \rangle = \sum_j e^{tk_j} p_j . \tag{D.29}$$

(b) For a continuous random variable X, the moment generating function is:

$$\mathcal{G}_x(t) = \langle e^{tx} \rangle = \int_{-\infty}^{+\infty} e^{tx} f(x) \, dx . \tag{D.30}$$

Let us consider here only continuous random variables. By expanding the exponential e^{tx} in (D.30) into a MacLaurin series, and taking into account that the mean of a sum is the sum of the means (Appendix D.8), one gets:

$$\begin{aligned}
\mathcal{G}_x(t) &= \left\langle 1 + tx + \frac{t^2 x^2}{2!} + \frac{t^3 x^3}{3!} + \cdots \right\rangle \\
&= 1 + t \langle x \rangle + \frac{t^2 \langle x^2 \rangle}{2!} + \frac{t^3 \langle x^3 \rangle}{3!} + \cdots = \sum_{s=0}^{\infty} \frac{t^s \alpha_s}{s!},
\end{aligned} \tag{D.31}$$

where α_s is the initial moment of order s of the random variable.

Let us now calculate the derivatives of the moment generating function with respect to t:

$$\frac{d^s \mathcal{G}_x(t)}{dt^s} = \frac{d^s \langle e^{tx} \rangle}{dt^s} = \langle x^s e^{tx} \rangle . \tag{D.32}$$

One can easily verify that the sth derivative of the moment generating function, calculated for $t = 0$, is the sth initial moment of the distribution of random variable:

$$\left. \frac{d^s \mathcal{G}_x(t)}{dt^s} \right|_{t=0} = \langle x^s \rangle = \alpha_s . \tag{D.33}$$

The same conclusion can be drawn for discrete random variables.

According to (D.33), the moments α_s can be calculated from the moment generating function. Some examples are encountered in the next sections.

A uniqueness theorem states that for two given random variables X and Y the equality of the moment generating functions, $\mathcal{G}_x(t) = \mathcal{G}_y(t)$, is a necessary and sufficient condition for the equality of the distribution functions, $f_x(x) = f_y(y)$.

It is worth remembering that the moment generating function does not necessarily exist for every distribution. For example, the Cauchy distribution (Sect. 6.7) has no moment generating function.

Characteristic Function

By substituting the real variable t in (D.29) or (D.30) with the imaginary variable $i\omega$ (where $i^2 = -1$), one obtains the *characteristic function* $\Psi(\omega)$. The characteristic function of a continuous random variable is

$$\Psi(\omega) = \langle e^{i\omega x} \rangle = \int_{-\infty}^{+\infty} e^{i\omega x} f(x)\,dx; \qquad (D.34)$$

the extension to a discrete random variable is trivial.

By expanding $e^{i\omega x}$ in (D.34), one obtains

$$\begin{aligned}\Psi(\omega) &= \left\langle 1 + i\omega x + \frac{(i\omega x)^2}{2!} + \frac{(i\omega x)^3}{3!} + \cdots \right\rangle \\ &= 1 + i\omega \langle x \rangle + \frac{(i\omega)^2 \langle x^2 \rangle}{2!} + \frac{(i\omega)^3 \langle x^3 \rangle}{3!} + \cdots \\ &= \sum_{s=0}^{\infty} \frac{(i\omega)^s \alpha_s}{s!}, \end{aligned} \qquad (D.35)$$

where α_s is the initial moment of order s.

A relation analogous to (D.33) holds for characteristic functions:

$$\alpha_s = \langle x^s \rangle = (-1)^s\, i^s\, \left.\frac{d^s\Psi(\omega)}{d\omega^s}\right|_{\omega=0}. \qquad (D.36)$$

A uniqueness theorem states that the equality of the characteristic functions, $\Psi_x(\omega) = \Psi_y(\omega)$, is a necessary and sufficient condition for the equality of the distribution functions, $f_x(x) = f_y(y)$.

The characteristic function has more general properties than the moment generating function. For example, a characteristic function exists for all distribution functions.

The functional relation connecting the variables X and ω in (D.34) is known as the *Fourier transform*.

Relations Between Initial and Central Moments

The *initial moments* α_s (Sect. 6.3) can be calculated from the moment generating function or the characteristic function through (D.33) or (D.36), respectively. The *central moments* μ_s can be calculated from the initial moments through simple relations; for the lowest-order moments,

$$\begin{aligned}\mu_2 &= \alpha_2 - \alpha_1^2, \\ \mu_3 &= \alpha_3 - 3\alpha_2\alpha_1 + 2\alpha_1^3, \\ \mu_4 &= \alpha_4 - 4\alpha_3\alpha_1 + 6\alpha_2\alpha_1^2 - 3\alpha_1^4. \end{aligned} \qquad (D.37)$$

D.3 Moments of the Binomial Distribution

Let us calculate the lowest-order moments of the binomial distribution (Sect. 6.1):

$$\mathcal{P}_{n,p}(k) = \binom{n}{k} p^k q^{n-k} = \frac{n!}{(n-k)!\,k!} p^k q^{n-k} \,. \tag{D.38}$$

Direct Calculation of Mean and Variance

Let us demonstrate that the mean of the binomial distribution is $m = np$:

$$m = \sum_{k=0}^{n} k \frac{n!}{(n-k)!\,k!} p^k q^{n-k} = np \sum_{k=1}^{n} \frac{(n-1)!}{(n-k)!\,(k-1)!} p^{k-1} q^{n-k}$$

$$= np \sum_{s=0}^{n-1} \frac{(n-1)!}{(n-1-s)!\,s!} p^s q^{n-1-s} = np \,. \tag{D.39}$$

In the last line, the substitution $k - 1 = s$ has been made: the resulting sum is one, because it represents the normalization condition for the binomial distribution $\mathcal{P}_{n-1,p}(s)$.

Let us now calculate the variance as

$$D = \langle k^2 \rangle - \langle k \rangle^2 \,. \tag{D.40}$$

From (D.39), $\langle k \rangle^2 = m^2 = (np)^2$. Let us calculate $\langle k^2 \rangle$:

$$\langle k^2 \rangle = \sum_{k=0}^{n} k^2 \frac{n!}{(n-k)!\,k!} p^k q^{n-k} = np \sum_{k=1}^{n} k \frac{(n-1)!}{(n-k)!\,(k-1)!} p^{k-1} q^{n-k}$$

$$= np \sum_{s=0}^{n-1} (s+1) \frac{(n-1)!}{(n-1-s)!\,s!} p^s q^{n-1-s}$$

$$= np\,[(n-1)p + 1] = (np)^2 - np^2 + np \,. \tag{D.41}$$

The last sum is the mean of $s + 1$ for the binomial distribution $\mathcal{P}_{n-1,p}(s)$: $\langle s+1 \rangle = \langle s \rangle + 1 = (n-1)\,p + 1$. One can easily conclude that

$$D = \langle k^2 \rangle - \langle k \rangle^2 = npq \,. \tag{D.42}$$

Calculation of the Moments from the Transformed Functions

The moment generating function of the binomial distribution is (Fig. D.1)

$$\mathcal{G}(t) = \langle e^{tk} \rangle = \sum_{k=0}^{n} e^{tk} \binom{n}{k} p^k q^{n-k}$$

$$= \sum_{k=0}^{n} \binom{n}{k} (pe^t)^k q^{n-k} = (pe^t + q)^n \,. \tag{D.43}$$

From (D.33), the first initial moments of the binomial distribution are:

$$\alpha_1 = np,\tag{D.44}$$
$$\alpha_2 = n(n-1)p^2 + np,$$
$$\alpha_3 = n(n-1)(n-2)p^3 + 3n(n-1)p^2 + np,$$
$$\alpha_4 = n(n-1)(n-2)(n-3)p^4 + 6n(n-1)(n-2)p^3 + 7n(n-1)p^2 + np,$$

and, from (D.37), the first central moments are:

$$\begin{array}{ll} \mu_1 = 0, & \mu_2 = npq, \\ \mu_3 = npq\,(q-p), & \mu_4 = npq\,[1 + 3npq - 6pq]. \end{array} \tag{D.45}$$

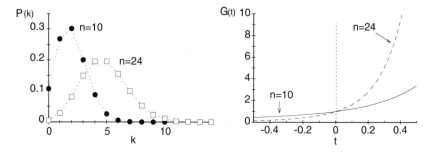

Fig. D.1. Binomial distributions for $p = 0.2$ and two different values of n (left) and the corresponding moment generating functions plotted around $t = 0$ (right).

The same results can be obtained from the characteristic function of the binomial distribution

$$\Psi(\omega) = \left\langle e^{i\omega k} \right\rangle = \sum_{k=0}^{n} e^{i\omega k} \binom{n}{k} p^k q^{n-k} = (pe^{i\omega} + q)^n,\tag{D.46}$$

using (D.36) for calculating the initial moments.

D.4 Moments of the Uniform Distribution

Let us consider the uniform distribution (Sects. 6.2 and 6.3):

$$f(x) = \begin{cases} 0 & \text{for } x < x_1 \\ C & \text{for } x_1 \leq x < x_2 \\ 0 & \text{for } x_2 \leq x \end{cases} \quad \left[C = \frac{1}{x_2 - x_1} \right]. \tag{D.47}$$

The mean is

$$m = \int_{-\infty}^{+\infty} x\, f(x)\, dx = \frac{1}{x_2 - x_1} \int_{x_1}^{x_2} x\, dx = \frac{x_1 + x_2}{2} . \tag{D.48}$$

The variance $D = \mu_2$ and the higher-order central moments $\mu_3, \mu_4, \ldots$ depend on the width and shape of the distribution, not on its position. To simplify the calculation, it is convenient to shift the distribution so that the mean is zero, $m = 0$. The distribution is then symmetric with respect to $x = 0$, and $x_1 = -\gamma, x_2 = +\gamma$, where γ is the half-width. The initial moments of the shifted distribution are calculated as

$$\alpha_s = \int_{-\infty}^{+\infty} x^s f(x)\, dx = \frac{1}{2\gamma} \int_{-\gamma}^{+\gamma} x^s\, dx , \tag{D.49}$$

and through (D.37), where now $m = 0$, one obtains the central moments

$$\mu_1 = 0, \quad \mu_2 = \gamma^2/3, \quad \mu_3 = 0, \quad \mu_4 = \gamma^4/5 , \tag{D.50}$$

that can be conveniently expressed as a function of the width $\Delta x = 2\gamma$:

$$\mu_1 = 0, \quad \mu_2 = (\Delta x)^2/12, \quad \mu_3 = 0, \quad \mu_4 = (\Delta x)^4/80 . \tag{D.51}$$

D.5 Moments of the Poisson Distribution

Let us calculate the lowest-order moments of the Poisson distribution (Sect. 6.4):

$$P_a(k) = \frac{a^k}{k!}\, e^{-a} . \tag{D.52}$$

Direct Calculation of Mean and Variance

The mean of the Poisson distribution is

$$m = \langle k \rangle = \sum_{k=0}^{\infty} k\, \frac{a^k}{k!}\, e^{-a} = \sum_{k=1}^{\infty} k\, \frac{a^k}{k!}\, e^{-a}$$

$$= a\, e^{-a} \sum_{k=1}^{\infty} \frac{a^{k-1}}{(k-1)!} = a\, e^{-a} \sum_{s=0}^{\infty} \frac{a^s}{s!} = a . \tag{D.53}$$

In the last line, after the substitution $s = k - 1$, one has taken into account that $\sum_{s=0}^{\infty} a^s/s! = e^a$.

To evaluate the variance, let us first calculate

$$\langle k^2 \rangle = \sum_{k=0}^{\infty} k^2\, \frac{a^k}{k!}\, e^{-a} = a \sum_{k=1}^{\infty} k\, \frac{a^{k-1}}{(k-1)!}\, e^{-a}$$

$$= a \left[\sum_{k=1}^{\infty} (k-1)\, \frac{a^{k-1}}{(k-1)!}\, e^{-a} + \sum_{k=1}^{\infty} \frac{a^{k-1}}{(k-1)!}\, e^{-a} \right]$$

$$= a\, [a + 1] , \tag{D.54}$$

where the substitution $k = (k-1) + 1$ has been made.

The variance is thus

$$D = \langle k^2 \rangle - \langle k \rangle^2 = a^2 + a - a^2 = a \ . \tag{D.55}$$

Calculation of Moments from the Transformed Functions

The moment generating function of the Poisson distribution is

$$\mathcal{G}(t) = \langle e^{tk} \rangle = \sum_{k=0}^{\infty} e^{tk} \frac{a^k}{k!} e^{-a}$$

$$= e^{-a} \sum_{k=0}^{\infty} \frac{(ae^t)^k}{k!} = e^{-a} e^{ae^t} = e^{a(e^t - 1)} \ . \tag{D.56}$$

From (D.33) and (D.37), the first initial and central moments of the Poisson distribution are:

$$\begin{aligned}
\alpha_1 &= a \ , & \mu_1 &= 0 \ , \\
\alpha_2 &= a^2 + a \ , & \mu_2 &= a \ , \\
\alpha_3 &= a^3 + 3a + a \ , & \mu_3 &= a \ , \\
\alpha_4 &= a^4 + 6a^3 + 7a^2 + a \ , & \mu_4 &= 3a^2 + a \ .
\end{aligned} \tag{D.57}$$

The same results can be obtained from the characteristic function

$$\Psi(\omega) = \langle e^{i\omega k} \rangle = \sum_{k=0}^{\infty} e^{i\omega k} \frac{a^k}{k!} e^{-a} = e^{a(e^{i\omega} - 1)} \ , \tag{D.58}$$

using (D.36) for calculating the initial moments.

D.6 Moments of the Normal Distribution

To calculate the first moments of the normal distribution (Sect. 6.5)

$$f(x) = \frac{1}{\sigma\sqrt{2\pi}} \exp\left[-\frac{(x-m)^2}{2\sigma^2}\right] \ , \tag{D.59}$$

it is first convenient to demonstrate the Eulero–Poisson integral

$$\int_{-\infty}^{+\infty} e^{-x^2} \, dx = \sqrt{\pi} \ . \tag{D.60}$$

Calculation of the Eulero–Poisson Integral

The calculation of the Eulero–Poisson integral can be divided into three steps.

Step 1

Let us consider the function $\exp(-x^2 - y^2)$, defined within a domain C of the xy plane, represented by the circle of radius $R = (x^2 + y^2)^{1/2}$ centered at the origin (Fig. D.2, left). By transforming to polar coordinates r, θ, it is easy to calculate the following integral

$$I_c = \iint_C e^{-x^2-y^2} \, dx \, dy = \int_0^{2\pi} d\theta \int_0^R e^{-r^2} r \, dr = \pi \left[1 - e^{-R^2}\right]. \quad (D.61)$$

When $R \to \infty$, the domain C extends to the entire plane, and $I_c \to \pi$.

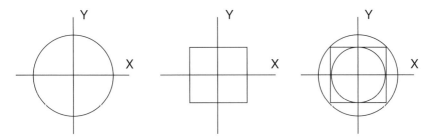

Fig. D.2. Graphical representation of the integration domains for the calculation of the Eulero–Poisson integral.

Step 2

Let us now consider a domain Q represented by a square of side $2a$, centered at the origin, $-a \le x \le +a$, $-a \le y \le +a$ (Fig. D.2, center). Our aim is now to calculate the integral

$$I_Q = \int_{-a}^{+a} dx \int_{-a}^{+a} dy \, e^{-x^2-y^2}. \quad (D.62)$$

To this aim, let us consider the two circles C_1 and C_2, of radii R_1 and R_2, inscribed and circumscribed, respectively, to the square (Fig. D.2, right); for the integrals calculated over the three domains C_1, Q, C_2, the following relation holds,

$$I_{C_1} < I_Q < I_{C_2}. \quad (D.63)$$

Let us now expand the square Q to the entire plane $(a \to \infty)$. Correspondingly, $R_1 \to \infty$, $R_2 \to \infty$, and, from (D.61), $I_{C_1} \to \pi$, $I_{C_2} \to \pi$. As a consequence, if $a \to \infty$, also $I_Q \to \pi$.

Step 3

It is easy to show that

$$\left[\int_{-a}^{+a} e^{-x^2} dx\right]^2 = \int_{-a}^{+a} dx \int_{-a}^{+a} dy\, e^{-x^2-y^2}. \qquad (D.64)$$

As a consequence

$$\left[\int_{-\infty}^{+\infty} e^{-x^2} dx\right]^2 = \lim_{a \to \infty} \left[\int_{-a}^{+a} e^{-x^2} dx\right]^2 = \pi, \qquad (D.65)$$

and (D.60) is demonstrated.

Direct Calculation of Moments

To calculate the moments of the normal distribution, it is convenient to consider the new variable

$$t = \frac{x-m}{\sigma\sqrt{2}} \quad \Rightarrow \quad x = \sigma\sqrt{2}\,t + m, \quad dx = \sigma\sqrt{2}\,dt. \qquad (D.66)$$

Let us first calculate the mean

$$\langle x \rangle = \int_{-\infty}^{+\infty} x f(x)\,dx = \frac{1}{\sqrt{\pi}} \int_{-\infty}^{+\infty} \left(\sigma\sqrt{2}\,t + m\right) e^{-t^2} dt$$

$$= \frac{\sigma\sqrt{2}}{\sqrt{\pi}} \int_{-\infty}^{+\infty} t\,e^{-t^2} dt + \frac{m}{\sqrt{\pi}} \int_{-\infty}^{+\infty} e^{-t^2} dt = m. \qquad (D.67)$$

In the last line, the first integral is zero, because the integrand is an odd function, and the second one is the Eulero-Posson integral (D.60).

Let us now demonstrate the recurrence relation

$$\mu_s = (s-1)\,\sigma^2\,\mu_{s-2}, \qquad (D.68)$$

that allows the evaluation of a central moment μ_s, once the moment μ_{s-2} is known. The central moment of order s is, by definition,

$$\mu_s = \int_{-\infty}^{+\infty} (x-m)^s f(x)\,dx = \frac{(\sigma\sqrt{2})^s}{\sqrt{\pi}} \int_{-\infty}^{+\infty} t^s e^{-t^2} dt. \qquad (D.69)$$

The integral in (D.69) can be calculated by parts

$$\int_{-\infty}^{+\infty} t^s e^{-t^2} dt = \underbrace{-\frac{1}{2} e^{-t^2} t^{s-1} \Big|_{-\infty}^{+\infty}}_{=0} + \frac{s-1}{2} \int_{-\infty}^{+\infty} t^{s-2} e^{-t^2} dt, \qquad (D.70)$$

D.6 Moments of the Normal Distribution

so that

$$\mu_s = \frac{(s-1)(\sigma\sqrt{2})^s}{2\sqrt{\pi}} \int_{-\infty}^{+\infty} t^{s-2} e^{-t^2} \, dt \, . \tag{D.71}$$

By substituting $s \to s-2$ in (D.69), one finds the expression for the central moment of order $s-2$,

$$\mu_{s-2} = \frac{(\sigma\sqrt{2})^{s-2}}{\sqrt{\pi}} \int_{-\infty}^{+\infty} t^{s-2} e^{-t^2} \, dt \, . \tag{D.72}$$

By comparing (D.71) and (D.72), one recovers the recurrence relation (D.68). Starting from the lowest order moments

$$\begin{aligned} \mu_0 &= 1 \quad \text{(normalization integral),} \\ \mu_1 &= 0 \quad \text{(average deviation from the mean),} \end{aligned} \tag{D.73}$$

all central moments can be calculated by the recurrence relation (D.68).
All central moments of odd order are zero:

$$\mu_3 = \mu_5 = \mu_7 = \cdots = 0 \, . \tag{D.74}$$

The values of the first even central moments are:

$$\mu_2 = \sigma^2, \quad \mu_4 = 3\sigma^4, \quad \mu_6 = 15\sigma^6 \, . \tag{D.75}$$

Moment Generating Function and Characteristic Function

The moment generating function of the normal distribution is

$$\mathcal{G}_x(t) = e^{mt} e^{\sigma^2 t^2/2} \, . \tag{D.76}$$

For the standard normal distribution

$$\phi(z) = \frac{1}{\sqrt{2\pi}} e^{-z^2/2} \, , \tag{D.77}$$

the moment generating function is particularly simple:

$$\mathcal{G}_z(t) = e^{t^2/2} \, . \tag{D.78}$$

The characteristic function of the normal distribution is

$$\Psi_x(\omega) = e^{im\omega} e^{-\sigma^2\omega^2/2} = [\cos(m\omega) + i\sin(m\omega)] \, e^{-\sigma^2\omega^2/2} \, . \tag{D.79}$$

For the standard normal distribution (D.77), the characteristic function is also particularly simple:

$$\Psi_z(\omega) = e^{-\omega^2/2} \, . \tag{D.80}$$

D.7 Parameters of the Cauchy Distribution

Let us first verify the normalization condition of the Cauchy distribution (Sect. 6.7):

$$\frac{1}{\pi} \int_{-\infty}^{+\infty} \frac{\gamma}{(x-\mu)^2 + \gamma^2} \, dx = 1. \tag{D.81}$$

By substituting

$$t = \frac{x-\mu}{\gamma} \quad \Rightarrow \quad x - \mu = \gamma t, \quad dx = \gamma \, dt, \tag{D.82}$$

one can easily find

$$\frac{1}{\pi} \int_{-\infty}^{+\infty} \frac{1}{t^2 + 1} dt = \frac{1}{\pi} \left[\lim_{a \to -\infty} \operatorname{arctg} t \Big|_a^0 \right] + \frac{1}{\pi} \left[\lim_{b \to +\infty} \operatorname{arctg} t \Big|_0^b \right] = 1. \tag{D.83}$$

Let us now try to calculate the mean, making use of the substitution (D.82).

$$\begin{aligned} \langle x \rangle &= \frac{1}{\pi} \int_{-\infty}^{+\infty} x \frac{\gamma}{(x-\mu)^2 + \gamma^2} dx \\ &= \frac{\gamma}{\pi} \int_{-\infty}^{+\infty} \frac{t}{t^2 + 1} dt + \frac{\mu}{\pi} \int_{-\infty}^{+\infty} \frac{1}{t^2 + 1} dt \,. \end{aligned} \tag{D.84}$$

In the last line, the second term is μ, whereas the first term

$$\frac{\gamma}{2\pi} \left[\lim_{a \to -\infty} \ln(t^2 + 1) \Big|_a^0 \right] + \frac{\gamma}{2\pi} \left[\lim_{b \to +\infty} \ln(t^2 + 1) \Big|_0^b \right] \tag{D.85}$$

is an undetermined form $-\infty + \infty$. The mean of the Cauchy distribution is not defined.

Finally, let us try to calculate the variance, making use of (D.82).

$$\begin{aligned} D_x &= \frac{1}{\pi} \int_{-\infty}^{+\infty} (x-\mu)^2 \frac{\gamma}{(x-\mu)^2 + \gamma^2} dx \\ &= \frac{\gamma^2}{\pi} \int_{-\infty}^{+\infty} \frac{t^2}{t^2 + 1} dt = \frac{\gamma^2}{\pi} \int_{-\infty}^{+\infty} \frac{t^2 + 1 - 1}{t^2 + 1} dt \\ &= \frac{\gamma^2}{\pi} \int_{-\infty}^{+\infty} dt - \frac{\gamma^2}{\pi} \int_{-\infty}^{+\infty} \frac{1}{t^2 + 1} dt \,. \end{aligned} \tag{D.86}$$

The first integral of the last line is divergent. The variance of the Cauchy distribution is not defined.

D.8 Theorems on Means and Variances

In this appendix, the relations

$$\mathbf{m}[aX] = a\,\mathbf{m}[X]\,, \tag{D.87}$$
$$\mathbf{D}[aX] = a^2\,\mathbf{D}[X]\,, \tag{D.88}$$
$$\mathbf{m}[X+Y] = \mathbf{m}[X] + \mathbf{m}[Y]\,, \tag{D.89}$$
$$\mathbf{D}[X+Y] = \mathbf{D}[X] + \mathbf{D}[Y]\,, \tag{D.90}$$

introduced in Sect. 7.2, are demonstrated.

Here, X and Y are two continuous random variables, and a a real constant. The symbols $\mathbf{m}[\ldots]$ and $\mathbf{D}[\ldots]$ label mean and variance, respectively. Equation (D.90) is true only for independent variables. The extension to discrete random variables, as well as to more than two variables (as required in Sect. 8.2), is straightforward.

(a) Demonstration of (D.87)

$$\mathbf{m}[aX] = \int_{-\infty}^{+\infty} a\,x\,f(x)\,\mathrm{d}x$$
$$= a \int_{-\infty}^{+\infty} x\,f(x)\,\mathrm{d}x = a\,\mathbf{m}[X] = am_x\,. \tag{D.91}$$

(b) Demonstration of (D.88)

$$\mathbf{D}[aX] = \int_{-\infty}^{+\infty} (ax - am_x)^2\,f(x)\,\mathrm{d}x$$
$$= a^2 \int_{-\infty}^{+\infty} (x - m_x)^2\,f(x)\,\mathrm{d}x = a^2\,\mathbf{D}[X] = a^2 D_x\,. \tag{D.92}$$

(c) Demonstration of (D.89)

Making use of the definition of the mean of a multivariate distribution (Sect. 6.8):

$$\mathbf{m}[X+Y] = \iint_{-\infty}^{+\infty} (x+y)\,f(x,y)\,\mathrm{d}x\,\mathrm{d}y$$
$$= \iint_{-\infty}^{+\infty} x\,f(x,y)\,\mathrm{d}x\,\mathrm{d}y + \iint_{-\infty}^{+\infty} y\,f(x,y)\,\mathrm{d}x\,\mathrm{d}y$$
$$= \mathbf{m}[X] + \mathbf{m}[Y] = m_x + m_y\,. \tag{D.93}$$

(d) Demonstration of (D.90)

Making use of the definitions of variance and covariance of a multivariate distribution (Sect. 6.8):

$$\begin{aligned}
&\mathbf{D}[X+Y] \\
&= \iint_{-\infty}^{+\infty} \left[(x+y) - (m_x + m_y)\right]^2 \, f(x,y) \, \mathrm{d}x \, \mathrm{d}y \\
&= \iint_{-\infty}^{+\infty} \left[(x - m_x) + (y - m_y)\right]^2 \, f(x,y) \, \mathrm{d}x \, \mathrm{d}y \\
&= \iint_{-\infty}^{+\infty} \left[(x - m_x)^2 + 2(x - m_x)(y - m_y) + (y - m_y)^2\right] f(x,y) \, \mathrm{d}x \, \mathrm{d}y \\
&= \mathbf{D}[X] + 2\sigma[XY] + \mathbf{D}[Y] = D_x + 2\sigma_{xy} + D_y \, .
\end{aligned} \qquad (\mathrm{D}.94)$$

In the last line of (D.94), $\sigma[XY] \equiv \sigma_{xy}$ is the covariance of the two random variables X and Y. Only if the two variables are independent, is the covariance equal to zero, and does (D.94) reduce to (D.90).

E Experiments

In this appendix, some simple laboratory experiments are presented. The main aim of the experiments is to give the reader an opportunity of exercising his or her skills on the procedures of data analysis.

E.1 Caliper and Micrometer

Aims of the Experiment

- Measurement of small lengths by instruments of different resolution
- Introduction to the uncertainty due to finite resolution (Sect. 4.2)
- Use of histograms and their normalization (Appendix A.4)
- Calculation of mean, variance, and standard deviation (Sect. 4.3)
- Practice on significant digits and rounding (Appendix A.1)

Materials and Instruments

- N objects produced in series, with length approximately 2 cm (cylinders, screws, bolts, etc.)
- A vernier caliper with resolution 0.05 mm (Fig. E.1, left)
- A micrometer with resolution 0.01 mm (Fig. E.1, right)

Introduction

The direct measurement of a physical quantity $\mathcal{G}$ is made by comparison with a unit standard $\mathcal{U}$. The result of a measurement is an interval of possible values, $n\mathcal{U} < X < (n+1)\mathcal{U}$, whose width $\Delta X = \mathcal{U}$ corresponds to the instrument resolution (Sect. 1.3). The finite resolution ΔX gives rise to an uncertainty δX, and the measurement result is expressed as $X = X_0 \pm \delta X$, where X_0 is the central value and $\delta X = \Delta X/2$ the maximum uncertainty (Sect. 4.2). The uncertainty δX can be reduced by reducing the resolution ΔX. Caliper and micrometer have different resolutions, $\Delta X = 0.05$ mm and 0.01 mm, respectively.

The N objects produced in series have the same nominal length. However, if the instrument has small enough resolution, it is possible to distinguish some differences between the different objects. As a consequence, the lengths of the N objects will not be equal to a unique value $X_0 \pm \delta X$, but will give rise to a distribution of values that can be conveniently represented by a histogram (Appendix A.4). The main characteristics of the distribution can be synthetically described by two parameters, mean and standard deviation (Sect. 6.3).

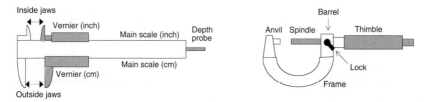

Fig. E.1. Schematic representation of a caliper (left) and a micrometer (right).

Experiment

0. Preliminary note. It is good practice that each step of an experiment be synthetically but exhaustively described in a laboratory logbook. One begins by writing the date and the name of the persons involved. All relevant information has to be carefully noted (instruments, numerical results, encountered difficulties, and so on).

1. Measurements with the caliper (A). Measure the length of the N objects with the vernier caliper (resolution 0.05 mm) and draw the corresponding histogram (A). Each bin of the histogram has a width proportional to the resolution $\Delta x_A = 0.05$ mm and a height proportional to the number $n^*_{j,A}$ of measures falling within the resolution interval. Here, the index i labels the measures ($i = 1, \ldots, N$), while j labels the histogram bins.

2. Measurements with the micrometer (B). Repeat the measurements using the micrometer (resolution 0.01 mm) and draw the corresponding histogram (B). The bin width and height are now proportional to $\Delta x_B = 0.01$ mm and $n^*_{j,B}$, respectively.

3. Height-normalized histograms. Normalize both A and B histograms to total unit height, by plotting on the vertical axis the sample frequencies

$$p^*_j = n^*_j/N \;. \tag{E.1}$$

4. Area-normalized histograms. Normalize both histograms to total unit area, by plotting on the vertical axis the sample densities

$$f^*_j = \frac{n_j}{N\,\Delta x} \;. \tag{E.2}$$

5. *Statistical parameters.* For both sets A and B of measurements, calculate the sample mean

$$m^* = \langle x \rangle = \frac{1}{N} \sum_{i=1}^{N} x_i = \sum_{j=1}^{\mathcal{N}} x_j \, p_j^*, \qquad (E.3)$$

the sample variance

$$D^* = \langle (x - m^*)^2 \rangle = \frac{1}{N} \sum_{i=1}^{N} (x_i - m^*)^2 = \sum_{j=1}^{\mathcal{N}} (x_j - m^*)^2 \, p_j^*, \qquad (E.4)$$

and the sample standard deviation

$$\sigma^* = \sqrt{D^*}. \qquad (E.5)$$

In (E.3) through (E.5), N is the total number of measurement values, and $\mathcal{N}$ is the number of histogram bins.

6. *Comparisons.* Compare the A and B histograms in the three versions (original, height-normalized, and area-normalized) and evaluate which version is the most suited for the comparison. Compare the means m^* and the standard deviations σ^* of the two measurement sets A and B, and discuss the origin of the possible differences between m_A^* and m_B^* and between σ_A^* and σ_B^*.

Discussion

7. *Parent and sample populations.* The N objects available are a sample of a larger population, made up of the batch of **N** objects produced by the factory (parent population). By convention, we use the asterisk * to label the sample parameters (such as sample mean and sample variance). Symbols without an asterisk represent the parameters of the parent population.

8. *Significant digits and approximations.* The results of measurements are always approximate values, because of their uncertainty. It is thus necessary to pay attention to the correct use of significant digits (Appendix A.1). When calculations are performed on approximate numerical values – e.g., to find means (E.3), variances (E.4), and standard deviations (E.5) – not all digits of the result are necessarily significant. The result has then to be rounded off in order to maintain only the significant digits, according to the rules given in Appendix A.1.

9. *Maximum uncertainty and standard uncertainty.* If the uncertainty δX is only due to the finite resolution ΔX, it is reasonable to assume $\delta X = \Delta X/2$. To facilitate the comparison with the uncertainties due to other causes, as well as for more subtle mathematical reasons, it is preferable to assume $\delta X = \Delta X/\sqrt{12}$ for the uncertainty due to resolution (Sect. 4.5). One thus distinguishes the maximum uncertainty $\delta X_{\mathrm{max}} = \Delta X/2$ from the standard uncertainty due to resolution $\delta X_{\mathrm{res}} = \Delta X/\sqrt{12}$.

E.2 Simple Pendulum: Measurement of Period

Aims of the Experiment

- Evaluation of random fluctuations in repeated measurements (Sect. 4.3)
- Study of sample distributions: histograms, means and standard deviations, distributions of mean values (Sect. 4.3)
- Introduction to the limiting distributions and to the estimation of their parameters; the normal distribution (Sects. 6.5 and 7.3)
- Comparison of different measurement methodologies

Materials and Instruments

- A pendulum (a suspended body)
- A tape-line (resolution 1 mm)
- A manual stopwatch (resolution 0.01 s)

Introduction

A pendulum is an object that can oscillate around a fixed suspension point. The period T is the time required for one complete oscillation. In a *simple pendulum*, the oscillating body has negligible size and is suspended through a string of negligible mass. The period T depends on the string length and on the oscillation amplitude. The dependence on amplitude is negligible for sufficiently small amplitudes (Fig. A.1 of Appendix A.3).

In this experiment, the period T (a constant quantity) is measured many times, obtaining different values because of random fluctuations (Sect. 4.3). The distribution of measures is represented by a histogram (Sect. A.3), or, more synthetically, by two parameters, mean and standard deviation.

The histogram shape has a random character. However, when the number N of measurements increases, its shape tends to stabilize, and mean and standard deviation tend to reduce their fluctuations. These observations lead to the concept of limiting distribution, which can often be modeled by the normal distribution (Sect. 6.5).

Experiment

1. Distribution of measures. Adjust the string length at about 90 cm, and make sure that the pendulum oscillates in a vertical plane with amplitude of about 10 degrees. Measure the period T by the stopwatch.

Repeat the measurement $N = 200$ times. Divide the set of $N = 200$ values into $M = 10$ subsets of $n = 20$ values. Draw and compare two histograms, one for one of the subsets of $n = 20$ values, the other for the total set of $N = 200$ values. The width of the histogram bins corresponds to the resolution $\Delta T = 0.01$ s. Normalize the histograms according to what you think is the most suitable.

For both histograms, calculate and compare the sample mean and the sample standard deviation; for the subset of $n = 20$ values:

$$m^*[T] = \frac{1}{n}\sum_{i=1}^{n} T_i , \quad \sigma^*[T] = \sqrt{\frac{1}{n}\sum_{i=1}^{n}(T_i - m^*)^2} . \tag{E.6}$$

For the set of $N = 200$ values, substitute n with N in (E.6). Express mean and standard deviation with the right number of significant digits (Appendix A.1).

2. *Distribution of sample means.* Calculate the sample mean $m_k^*[T]$ and the standard deviation $\sigma_k^*[T]$ ($k = 1, \ldots, M$) (E.6) for each of the $M = 10$ subsets of $n = 20$ measures. Draw the histogram of the sample means m_k^*, and calculate its sample mean $m^*[m^*]$ and sample standard deviation $\sigma^*[m^*]$.

3. *Limiting distribution.* When the number N of measures increases, the histogram tends to a bell shape (Sect. 4.3), which can be analytically described by a normal distribution (Sect. 6.5),

$$f(x) = \frac{1}{\sigma\sqrt{2\pi}} \exp\left[-\frac{(x-m)^2}{2\sigma^2}\right], \tag{E.7}$$

where in this case $x \equiv T$. The values of the mean m and of the standard deviation σ of the limiting distribution (E.7) cannot be exactly calculated from a finite number of measures; they can only be estimated. The best estimate $\tilde{m}$ of the mean m is the sample mean m^*, and the best estimate $\tilde{\sigma}$ of the standard deviation σ is the sample standard deviation σ^* multiplied by $[N/(N-1)]^{1/2}$ (Sect. 7.3).

Estimate the parameters m and σ of the limiting distribution of the $N = 200$ measures:

$$\tilde{m}[T] = m^*[T], \quad \tilde{\sigma}[T] = \sqrt{\frac{N}{N-1}}\,\sigma^*[T], \tag{E.8}$$

and compare the corresponding normal distribution (E.7) with the area-normalized histogram of the $N = 200$ values.

Estimate the parameters m and σ of the limiting distribution of the sample means m^* of the subsets of $n = 20$ measurements, from the sample of $M = 10$ values m_k^*:

$$\tilde{m}[m^*] = m^*[m^*], \quad \tilde{\sigma}[m^*] = \sqrt{\frac{M}{M-1}}\,\sigma^*[m^*], \tag{E.9}$$

and compare the corresponding normal distribution (E.7) with the area-normalized histogram of the $M = 10$ values.

Compare the means and the standard deviations estimated for the two limiting distributions, of single measures and of sample means m^* of $n = 20$ values. Calculate the sample ratio $r^* = \tilde{\sigma}[T]/\tilde{\sigma}[m^*]$, and compare with the expected ratio for the parameters of the parent population, $r = \sigma[T]/\sigma[m^*] = n^{1/2}$ (Sect. 7.2).

4. *Uncertainty due to random fluctuations.* Synthesize the result of the repeated measurement of the period as

$$\mathcal{T} = \mathcal{T}_0 \pm \delta \mathcal{T}, \quad \text{(central value} \pm \text{uncertainty)}. \tag{E.10}$$

Assume as central value $\mathcal{T}_0$ the mean m of the limiting distribution, whose best estimate is the sample mean m^* (Sect. 4.3):

$$\mathcal{T}_0 = m^* = \frac{1}{N} \sum_{i=1}^{N} \mathcal{T}_i. \tag{E.11}$$

The uncertainty $\delta\mathcal{T}_{\text{cas}}$ depends on the fluctuations of the sample mean m^* (experimental estimate of $\mathcal{T}_0$) with respect to the unknown population mean m (assumed as the true value of $\mathcal{T}$). A measure of the fluctuations of m^* with respect to m is given by the standard deviation of the distribution of sample means. By convention, we assume $\delta\mathcal{T}_{\text{cas}} = \sigma[m^*]$ (standard uncertainty).

The standard deviations of the distributions of sample means and of single values are connected by $\sigma[m^*] = \sigma[\mathcal{T}]/N^{1/2}$. In turn, $\sigma[\mathcal{T}]$ can be estimated from the sample standard deviation σ^* as in (E.8). As a consequence the uncertainty is calculated as

$$\begin{aligned}\delta\mathcal{T}_{\text{cas}} &= \sigma[m^*] = \frac{1}{\sqrt{N}}\sigma = \frac{1}{\sqrt{N}}\sqrt{\frac{N}{N-1}}\sigma^* \\ &= \sqrt{\frac{1}{N(N-1)} \sum_{i=1}^{N} (\mathcal{T}_i - m^*)^2}.\end{aligned} \tag{E.12}$$

The uncertainty $\delta\mathcal{T}_{\text{cas}}$ decreases when N increases.

5. *Instrument resolution and measurement resolution.* The resolution of single measurements corresponds to the stopwatch resolution $\Delta\mathcal{T} = 0.01\,\text{s}$. The resolution can be reduced by measuring the sum of many periods. For example, one can measure the duration T_{10} of 10 oscillations and calculate the period as $\mathcal{T} = T_{10}/10$. The instrument resolution now refers to the value T_{10}, and a resolution ten times smaller, $\Delta\mathcal{T} = 0.001\,\text{s}$, can be attributed to the period value.

Repeat the measurement $n = 20$ times with this new procedure, draw the corresponding histogram and calculate its mean and standard deviation. Critically compare the uncertainty values obtained by the two procedures.

6. *Systematic errors.* A possible cause of systematic errors (Sect. 4.4) is connected with the reaction time of the experimenter, and the possible different behavior in starting and stopping the stopwatch. To evaluate this effect, compare the results obtained by different experimenters.

Discussion

7. Contributions to uncertainty. The experiment allows a comparison of the three contributions to uncertainty: (a) resolution, (b) random fluctuations, and (c) evaluation of systematic errors (Chap. 4). The uncertainty δT_{cas} due to random fluctuations (E.12) decreases when the number N of measures increases; it makes no sense, however, to reduce the uncertainty due to random fluctuations below the value of the uncertainty due to resolution δT_{res}. In addition, it can be meaningless to reduce the uncertainty due to random fluctuations and/or to resolution if one cannot guarantee that it is any larger than the uncertainty due to uncompensated systematic errors.

8. Sample and parent populations. The results of the N measurements can be considered a sample of a parent population, made up of all the possible (infinite) measurement results.

9. Standard expression of the uncertainty. The uncertainty δT_{cas} due to random fluctuations is measured by the standard deviation of the distribution of sample means, which includes about 68% of values m^*. No maximum uncertainty can be quoted, because the normal distribution is not bounded.

The uncertainty due to resolution, estimated as $\delta T_{\max} = \Delta T/2$, is instead a maximum uncertainty. To guarantee the possibility of comparing uncertainties due to different causes, the uncertainty due to resolution has also to be expressed as a standard uncertainty, corresponding to the standard deviation of the uniform distribution, $\delta T_{\text{res}} = \Delta T/\sqrt{12}$ (Sect. 4.5).

E.3 Helicoidal Spring: Elastic Constant

Aims of the Experiment

- Static measurement of the elastic constant of a spring
- Introduction to the propagation of uncertainty (Chap. 8)
- Introduction to the weighted average (Sect. 4.4)
- Introduction to linear graphs (Appendix A.3)
- Introduction to linear regression (Sect. 10.3) and chi square test (Chap. 11)

Materials and Instruments

- Several springs of different elastic constants
- Several bodies of different masses, to be suspended from the springs
- A mechanical balance (resolution 0.1 g)

Introduction

The deformation x of a spring is proportional to the applied force, $F = kx$, where k is the elastic constant (Fig E.2). In this experiment, the procedure by which the law $F = kx$ is inferred from the measurement of two physical quantities, force and deformation, is reproduced and critically analyzed. Notice that the elastic reaction of the spring is $F_e = -F = -kx$.

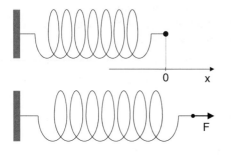

Fig. E.2. Helicoidal spring, undeformed (top) and deformed by a force $\boldsymbol{F}$ (bottom).

Experiment

1. Measurement of force. In this experiment, the springs are vertically suspended from a fixed point, and the force F is the weight P of the bodies suspended from the free end of the springs. The intensity of the force is varied by varying the total mass of the bodies.

First measure the masses m_i of the different groups of bodies that will be suspended from the spring, and express their uncertainty in the standard form $\delta m_i = \Delta m/\sqrt{12}$, where Δm is the resolution of the balance (Sect. 4.2). The weights are $P_i = m_i g$, where g is the acceleration of gravity (assume $g = 9.806\,\mathrm{m\,s^{-2}}$, and neglect its uncertainty). The uncertainty of the weights is $\delta P_i = g\,\delta m_i$.

2. Measurement of deformation. Choose one of the springs, and suspend it from one end. Suspend, from the other end of the spring, one of the bodies, wait for a complete damping of oscillations, and measure the spring elongation x_i. Progressively add the other bodies in order to increase the applied force, and measure the corresponding elongations. Evaluate the uncertainty of the elongation value that is probably due only to resolution, $\delta x = \Delta x/\sqrt{12}$ (Sect. 4.5).

3. Table and graph. List the values of force $P_i \pm \delta P_i$ and elongation $x_i \pm \delta x_i$ in a table (Sect. A.2). Plot the results in a graph (Sect. A.3), with P_i and x_i on the horizontal and vertical axis, respectively; plot the error bars. Can you decide, by visual inspection, whether the results are consistent with the direct proportionality relation $x = (1/k)P$?

4. *Calculation of the elastic constant from the table.* For each value P_i, calculate the ratio $k_i = P_i/x_i$, and evaluate its uncertainty by the propagation rule for quotients (Sect. 8.3):

$$\left(\frac{\delta k_i}{k_i}\right)^2 = \left(\frac{\delta P_i}{P_i}\right)^2 + \left(\frac{\delta x_i}{x_i}\right)^2. \tag{E.13}$$

Notice that the value of δk progressively decreases when the applied force increases. Besides, the two contributions $(\delta x_i/x_i)$ and $(\delta P_i/P_i)$ in (E.13) are quite different, and one of them could be neglected. To synthesize the $\mathcal{N}$ values $k_i \pm \delta k_i$ into one unique value $k_0 \pm \delta k$, two different procedures can be used.

(a) Calculate the weighted average of the values k_i and the corresponding uncertainty (Sect. 7.3):

$$k_0 = \frac{\sum_i k_i w_i}{\sum_i w_i}, \quad \delta k = \frac{1}{\sqrt{\sum_i w_i}}, \quad \text{where} \quad w_i = \frac{1}{(\delta k_i)^2}. \tag{E.14}$$

This procedure takes into account the uncertainties δk_i of the single values from (E.13), but does not account for random fluctuations.

(b) To take into account the possible influence of random fluctuations, consider the distribution of the values k_i, and calculate the sample mean m_k^* and the standard deviation of the distribution of sample means $\sigma[m_k^*]$. At last, according to Sect. 4.3,

$$k_0 = m_k^*, \quad \delta k = \sigma[m_k^*]. \tag{E.15}$$

Compare and discuss the results of the two procedures.

5. *Calculation of the elastic constant from the graph.* The elastic constant k can be obtained by measuring the slope $b = 1/k$ of the straight line that best fits the data points (P_i, x_i) (Sect. 10.3).

A first procedure, the graphical method, consists of drawing the straight lines of maximum and minimum slope compatible with the uncertainty crosses of the experimental points (Fig. 10.2 in Sect. 10.3). Their angular coefficients, measured by a goniometer, define an interval of values from which the slope $b_0 \pm \delta b$ can be evaluated.

A more accurate procedure is linear regression (Sect. 10.3). The best estimate of b is the value that minimizes the discrepancy between theory and experiment, expressed as

$$\chi^2 = \sum_{i=1}^{\mathcal{N}} \frac{(x_i - bP_i)^2}{(\delta x_i)^2}. \tag{E.16}$$

Because the uncertainties δx_i are all equal, one can show (Sect. 10.3) that

264   E Experiments

$$b_0 = \frac{\sum_i P_i x_i}{\sum_i P_i^2}, \quad \delta b = \frac{\delta x}{\sqrt{\sum_i P_i^2}}. \tag{E.17}$$

The uncertainty of the elastic constant k is obtained through the rule of propagation for the quotient (Sect. 8.3):

$$\delta k/k = \delta b/b. \tag{E.18}$$

Compare the values of k obtained by the different procedures (from the table and from the graph) and critically discuss the possible differences.

6. *The chi square test.* To evaluate the compatibility of the experimental values with the law of direct proportionality, make use of the chi square test (Chap. 11). The chi square test is based on the comparison of the discrepancies between theory and experiment with the uncertainties of the experimental points. The chi square has been already defined in (E.16):

$$\chi^2 = \sum_{i=1}^{\mathcal{N}} \frac{(x_i - b_0 P_i)^2}{(\delta x_i)^2}. \tag{E.19}$$

Notice that an approximate value of χ^2 can be obtained by comparing the discrepancy theory-experiment and the uncertainty of each point by simple visual inspection of the graph (Problem 11.1 of Chap. 11).

If the theory is correct, and the uncertainties have been correctly evaluated, one expects that, for each point, the discrepancy is comparable with the uncertainty, $(x_i - b_0 P_i)^2 \simeq (\delta x_i)^2$. The value b_0 has been evaluated by the linear regression procedure, therefore only $\mathcal{N} - 1$ experimental points are actually independent, so that one expects

$$\chi^2 = \sum_{i=1}^{\mathcal{N}} \frac{(x_i - b_0 P_i)^2}{(\delta x_i)^2} \simeq \mathcal{N} - 1. \tag{E.20}$$

The value of chi square can be qualitatively interpreted as follows.

(a) If $\chi^2 \simeq \mathcal{N} - 1$, and if the uncertainties are correctly evaluated, it is highly probable that the law $P = kx$ is consistent with the experimental points.
(b) If $\chi^2 \gg \mathcal{N} - 1$, the experimental data are not consistent with the law $P = kx$, or the uncertainties δx_i have been underevaluated.
(c) If $\chi^2 \ll \mathcal{N} - 1$, probably the uncertainties δx_i have been overevaluated.

In this experiment, the law $P = kx$ is known to be valid, and the χ^2 test can be used to check if the uncertainties have been correctly evaluated. If $\chi^2 \gg \mathcal{N} - 1$, one can evaluate an average uncertainty by inverting (E.20):

$$(\delta x)^2 \simeq \frac{1}{\mathcal{N} - 1} \sum_{i=1}^{\mathcal{N}} (x_i - b_0 P_i)^2. \tag{E.21}$$

A more refined probabilistic interpretation of the χ^2 test can be found in Sect. 11.4.

7. *Reproducibility of the experiment.* Progressively remove the bodies suspended from the spring, and measure the corresponding elongations. Repeat the procedure of loading and unloading the spring several times, and plot all the results in the same graph. Is the value of k dependent on the procedure of measurement (loading or unloading the spring)? Is the result of the experiment reproducible?

8. *Comparison of different springs.* Repeat the experiment with other springs, and plot the data on a graph. Evaluate and compare the values of the elastic constants of different springs.

Discussion

9. *Propagation of the uncertainty.* When a quantity Q is indirectly measured, $Q = f(X, Y, Z, \ldots)$, its uncertainty can be calculated by the general expression (Sect. 8.3):

$$(\delta Q)^2 \simeq \left(\frac{\partial Q}{\partial X}\right)_0^2 (\delta X)^2 + \left(\frac{\partial Q}{\partial Y}\right)_0^2 (\delta Y)^2 + \left(\frac{\partial Q}{\partial Z}\right)_0^2 (\delta Z)^2 + \cdots . \quad \text{(E.22)}$$

For sums $Q = X + Y$ and differences $Q = X - Y$, (E.22) becomes:

$$(\delta Q)^2 = (\delta X)^2 + (\delta Y)^2 . \quad \text{(E.23)}$$

For products $Q = XY$, and quotients $Q = X/Y$, from (E.22) one gets:

$$\left(\frac{\delta Q}{Q_0}\right)^2 \simeq \left(\frac{\delta X}{X_0}\right)^2 + \left(\frac{\delta Y}{Y_0}\right)^2 . \quad \text{(E.24)}$$

10. *Simple averages and weighted averages.* The simple average of two values x_a and x_b is

$$\langle x \rangle = \frac{x_a + x_b}{2} . \quad \text{(E.25)}$$

If the two values are weighted by w_a and w_b, respectively, their weighted average (Sect. 4.4) is

$$x_w = \frac{x_a w_a + x_b w_b}{w_a + w_b} . \quad \text{(E.26)}$$

If $w_a = w_b$, the weighted average (E.26) reduces to the simple average (E.25).

If x_a and x_b are measures affected by the uncertainties δx_a and δx_b, respectively, it is reasonable to assume the weights in (E.26) as

$$w_a = 1/(\delta x_a)^2 , \quad w_b = 1/(\delta x_b)^2 . \quad \text{(E.27)}$$

The uncertainty on the weighted average (Sect. 4.4) is

$$\delta x_w = \frac{1}{\sqrt{w_a + w_b}} . \quad \text{(E.28)}$$

Equations (E.26) and (E.28) can easily be generalized to more than two terms.

E.4 Helicoidal Spring: Oscillations

Aims of the Experiment

- Search for a relation among mass, elastic constant, and oscillation period of a spring
- Dynamic measurement of the elastic constant of a spring
- Introduction to nonlinear graphs (Appendix A.3)

Materials and Instruments

- Several suspended springs of different elastic constants
- Several bodies of different masses, to be suspended from the springs
- A stopwatch (resolution 0.01 s)
- A mechanical balance (resolution 0.1 g)

Introduction

A body of mass m, suspended from the free end of a spring of elastic constant k, periodically oscillates with respect to the equilibrium position. By comparing the behavior of different springs, one can qualitatively verify that the period of oscillation T increases when the mass m increases and/or the elastic constant k decreases. In this experiment, the quantitative relation among period T, elastic constant k, and mass m is determined.

Experiment

1. Dimensional analysis. Before performing any measurement, one can guess the form of the analytic relation among period, mass, and elastic constant by dimensional considerations (Sect. 2.5). Let us suppose that the period T can depend on mass m, elastic constant k, and oscillation amplitude A, according to a general relation such as $T \propto m^\alpha k^\beta A^\gamma$. The corresponding dimensional equation is

$$[T]^1 = [M]^{\alpha+\beta} [L]^\gamma [T]^{-2\beta}, \qquad (E.29)$$

which is satisfied for $\gamma = 0$, $\beta = -1/2$, $\alpha = -\beta = 1/2$, so that

$$T = C\sqrt{m/k}. \qquad (E.30)$$

The period does not depend on the amplitude of oscillations; the value of the constant C cannot be determined solely by dimensional considerations.

E.4 Helicoidal Spring: Oscillations

2. Dependence of the period on mass. Choose one of the springs and measure the periods corresponding to different suspended masses. Evaluate the uncertainties on the values of masses and periods. Plot the obtained values and the corresponding uncertainties on a graph, with masses and periods on the horizontal and vertical axis, respectively.

According to (E.30), the plotted points should not be consistent with a straight line. A linear appearance can be obtained by plotting on the vertical axis the values T^2. The uncertainty of T^2, according to the propagation rules of Sect. 8.3, is $\delta(T^2) = 2T\,\delta T$.

3. Effective mass of the spring. A spring has its own inertia, and can oscillate by itself, without being connected to another body. This effect can be appreciated in the graph of T^2 as a function of m: the straight line best fitting the experimental points crosses the horizontal axis in correspondence to a negative value $-m_e$. The value m_e can be interpreted as an effective mass, that measures the spring inertia. Equation (E.30) has to be modified, by substituting m with $M = m + m_e$, so that:

$$T^2 = \frac{C^2}{k} M = \frac{C^2}{k} m_e + \frac{C^2}{k} m, \qquad (\text{E.31})$$

where C and m_e are unknowns to be determined.

4. Linear regression. To determine the values of C and m_e, let us notice that (E.31) is a linear equation of the form

$$y = A + B\,x, \qquad (\text{E.32})$$

where $x \equiv m$, $y \equiv T^2$, $A \equiv C^2 m_e/k$, and $B \equiv C^2/k$. The parameters A and B in (E.32) can be estimated by linear regression (Sect. 10.3), say by minimizing the quantity

$$\chi^2 = \sum_{i=1}^{N} \frac{(y_i - A - Bx_i)^2}{(\delta y_i)^2}. \qquad (\text{E.33})$$

The values A and B that minimize (E.33) are

$$A = \frac{(\sum_i w_i x_i^2)(\sum_i w_i y_i) - (\sum_i w_i x_i)(\sum_i w_i x_i y_i)}{\Delta}, \qquad (\text{E.34})$$

$$B = \frac{(\sum_i w_i)(\sum_i w_i x_i y_i) - (\sum_i w_i y_i)(\sum_i w_i x_i)}{\Delta}, \qquad (\text{E.35})$$

where

$$\Delta = (\sum_i w_i)(\sum_i w_i x_i^2) - (\sum_i w_i x_i)^2, \qquad (\text{E.36})$$

and the weights w_i are connected to the uncertainties by

$$w_i = 1/(\delta y_i)^2. \qquad (\text{E.37})$$

The uncertainties of A and B, according to the propagation rules of Sect. 8.3, are

$$(\delta A)^2 = \frac{\sum_i w_i x_i^2}{\Delta}, \quad (\delta B)^2 = \frac{\sum_i w_i}{\Delta}. \tag{E.38}$$

From the values $A \pm \delta A$ and $B \pm \delta B$, one can recover the values C and m_e of (E.31), and their uncertainties, using the values of elastic constant k measured in the previous experiment E.3.

The above linear regression procedure requires that the uncertainties δx_i on the horizontal axis are negligible. If the uncertainties δx_i are not negligible, one can transfer the uncertainty of each value from the horizontal to the vertical axis, according to the procedure described in Sect. 10.4.

5. *The chi square test.* The chi square test (Chap. 11) allows us to evaluate the compatibility of the experimental values with the expected behavior (E.32). Because the two parameters A and B have been determined from the experimental points, there are $\mathcal{N} - 2$ degrees of freedom, so that one expects

$$\chi^2 = \sum_{i=1}^{\mathcal{N}} \frac{(y_i - A - Bx_i)^2}{(\delta y_i)^2} \simeq \mathcal{N} - 2. \tag{E.39}$$

(a) If $\chi^2 \simeq \mathcal{N} - 2$, and the uncertainties are correctly evaluated, the experimental data can be considered consistent with the law $y = A + Bx$.
(b) If $\chi^2 \gg \mathcal{N} - 2$, the experimental data are not consistent with the law $y = A + Bx$, or the uncertainties δy_i have been underevaluated.
(c) If $\chi^2 \ll \mathcal{N} - 2$, probably the uncertainties δy_i have been overevaluated.

6. *Dynamical determination of the elastic constant k.* According to the harmonic oscillator theory, the constant C in (E.30) has the exact value $C = 2\pi$, and the expression of the period is

$$T = 2\pi \sqrt{M/k}, \tag{E.40}$$

where $M = m + m_e$. By imposing $C = 2\pi$ in (E.31), one can now calculate k and m_e from the parameters A and B of the linear regression, and obtain a dynamical measurement of the elastic constant k. Is this value of k consistent with the value determined in the previous experiment E.3?

7. *Dependence of the period on the elastic constant.* Evaluate the relation between the elastic constant and oscillation period, by suspending the same body of mass m from springs of different elastic constant k (determined in the previous experiment E.3) and by measuring the corresponding period T. One can check the validity of (E.30) by plotting the values T^2 against $1/k$.

Discussion

8. *Effect of gravity on oscillations.* A body subject only to the elastic force, $F_e = -kx$, obeys the differential equation of motion

$$\frac{d^2x}{dt^2} = -\frac{k}{m}x, \qquad (E.41)$$

and oscillates around the equilibrium position $x = 0$. In this experiment, the spring is vertical, and the suspended body is subjected to both the elastic force and the gravity force, so that the equation of motion is

$$\frac{d^2x}{dt^2} = -\frac{k}{m}x + g. \qquad (E.42)$$

The elastic force and the gravity force are in equilibrium when $mg - kx = 0$, say $x = mg/k$. Let us substitute the new coordinate $y = x - mg/k$ in (E.42). The resulting equation,

$$\frac{d^2y}{dt^2} = -\frac{k}{m}y, \qquad (E.43)$$

is identical to (E.41); the period is thus unchanged, only the equilibrium position has changed from $x = 0$ to $x = mg/k$.

E.5 Simple Pendulum: Dependence of Period on Length

Aims of the Experiment

- Search for a relation between length and period
- Use of logarithmic graphs (Appendix A.3)
- Applications of linear regression (Sect. 10.3) and chi square test (Chap. 11)
- Measurement of the acceleration of gravity

Materials and Instruments

- A pendulum (a suspended body)
- A tape-line (resolution 1 mm)
- A caliper (resolution 0.05 mm)
- A stopwatch (resolution 0.01 s)

Introduction

A pendulum oscillates around its equilibrium position. It is easy to qualitatively verify that the period T of oscillation increases when the length ℓ of the pendulum increases. The first aim of this experiment is to determine the quantitative relation between period and length.

Within the approximation of small oscillations, the period T depends on length ℓ and acceleration of gravity g according to

$$T = 2\pi \sqrt{\ell/g}. \qquad (E.44)$$

The second aim of this experiment is to determine the value of the acceleration of gravity g from the measurements of length and period:

$$g = (2\pi/T)^2 \ell. \qquad (E.45)$$

Experiment

1. Measurement of lengths and periods. Measure the length of the pendulum, say the distance between the point of suspension of the string and the center of mass of the suspended body, and evaluate its uncertainty. Measure the period T and evaluate its uncertainty, by the same procedure as in Experiment E.2. Repeat the measurements of length and period for about ten different values of length, the minimum length being about 1/10 of the maximum length. Make sure that the pendulum oscillates within a vertical plane, and the amplitude of oscillations is constant.

2. Linear graph. Plot the N measures in a graph, length ℓ and period T on the horizontal and vertical axis, respectively, including the uncertainty bars.

Verify that the experimental points are inconsistent with a linear relation. Dimensional considerations (Sect. 2.5) lead to a proportionality relation between T and $\ell^{1/2}$. One could obtain a linear behavior by plotting T^2 against ℓ, as in experiment E.4. An alternative procedure is used here.

The experimental points appear to be consistent with a law

$$T = a\,\ell^b, \tag{E.46}$$

where a and b are unknown. To verify the soundness of (E.46) and determine the values of a and b, it is convenient to linearize the graph of experimental points by using logarithmic scales (Appendix A.3).

3. Logarithmic graph. Let us consider the new variables

$$X = \log(\ell), \quad Y = \log(T). \tag{E.47}$$

If (E.46) is right, the relation between Y and X is linear:

$$\log(T) = \log(a) + b\log(\ell), \quad \text{say } Y = A + BX, \tag{E.48}$$

where

$$A \equiv \log(a), \quad (a \equiv 10^A), \quad B \equiv b. \tag{E.49}$$

To verify the consistency of experimental data with (E.48), one can:

(a) Plot $Y = \log(T)$ against $X = \log(\ell)$ in a graph with linear scales, the uncertainties being calculated according to the propagation rules:

$$\delta Y = \left|\frac{\mathrm{d}Y}{\mathrm{d}T}\right|\delta T = \frac{\log(e)}{T}\delta T = \frac{0.43}{T}\delta T, \tag{E.50}$$

$$\delta X = \left|\frac{\mathrm{d}X}{\mathrm{d}\ell}\right|\delta\ell = \frac{\log(e)}{\ell}\delta\ell = \frac{0.43}{\ell}\delta\ell. \tag{E.51}$$

(b) Plot T against ℓ in a graph with logarithmic scales, the uncertainty bars corresponding to the original values $\delta\ell$ and δT.

E.5 Simple Pendulum: Dependence of Period on Length

4. Evaluation of A and B. The parameters $A \equiv \log(a)$ and $B \equiv b$ are the intercept and the angular coefficient, respectively, of the straight line best fitting the experimental points in the graph of $\log(T)$ against $\log(\ell)$.

As a first step, it is convenient to obtain rough values of A and B by the graphical method introduced in Sect. 10.3:

$$A = \log(T), \text{ for } \log(\ell) = 0\,; \quad b = B = \frac{\log(T_2) - \log(T_1)}{\log(\ell_2) - \log(\ell_1)}\,. \tag{E.52}$$

The value B from (E.52) allows us to transform the uncertainty of $X = \log(\ell)$ into a contribution to the uncertainty of $Y = \log(T)$ (Sect. 10.4),

$$(\delta Y_i)_{\text{tra}} = \left|\frac{dY}{dX}\right| \delta X_i = |B|\, \delta X_i\,, \tag{E.53}$$

to be quadratically added to the original uncertainty $(\delta Y_i)_{\text{exp}}$:

$$(\delta Y_i)^2_{\text{tot}} = (\delta Y_i)^2_{\text{exp}} + (\delta Y_i)^2_{\text{tra}}\,. \tag{E.54}$$

The parameters A and B of (E.48) can now be obtained by linear regression (Sect. 10.3), say by minimizing the quantity

$$\chi^2 = \sum_{i=1}^{N} \frac{(Y_i - A - BX_i)^2}{(\delta Y_i)^2_{\text{tot}}}\,. \tag{E.55}$$

The values A and B that minimize (E.55) are

$$A = \frac{(\sum_i w_i X_i^2)(\sum_i w_i Y_i) - (\sum_i w_i X_i)(\sum_i w_i X_i Y_i)}{\Delta}\,, \tag{E.56}$$

$$B \equiv b = \frac{(\sum_i w_i)(\sum_i w_i X_i Y_i) - (\sum_i w_i Y_i)(\sum_i w_i X_i)}{\Delta}\,, \tag{E.57}$$

where

$$\Delta = \left(\sum_i w_i\right)\left(\sum_i w_i X_i^2\right) - \left(\sum_i w_i X_i\right)^2\,, \tag{E.58}$$

and the weights w_i are connected to the uncertainties by

$$w_i = 1/(\delta Y_i)^2\,. \tag{E.59}$$

The uncertainties of A and B (Sect. 8.3) are

$$(\delta A)^2 = \frac{\sum_i w_i X_i^2}{\Delta}\,, \quad (\delta B)^2 \equiv (\delta b)^2 = \frac{\sum_i w_i}{\Delta}\,. \tag{E.60}$$

One can finally obtain the value $a \pm \delta a$:

$$a = 10^A\,, \quad \delta a = \left|\frac{da}{dA}\right| \delta A = 10^A \ln(10)\, \delta A = 2.3 \times 10^A\, \delta A\,. \tag{E.61}$$

5. *The chi square test.* The chi square test (Chap. 11) allows us to evaluate the compatibility of the experimental values with the expected behavior (E.48). Because the parameters A and B have been determined from the experimental points, there are $\mathcal{N} - 2$ degrees of freedom, and one expects

$$\chi^2 = \sum_{i=1}^{\mathcal{N}} \frac{(Y_i - A - BX_i)^2}{(\delta Y_i)^2} \simeq \mathcal{N} - 2 \,. \tag{E.62}$$

(a) If $\chi^2 \simeq \mathcal{N} - 2$, and the uncertainties are correctly evaluated, the experimental data can be considered consistent with the law $y = A + BX$.
(b) If $\chi^2 \gg \mathcal{N} - 2$, the experimental data are not consistent with the law $y = A + BX$, or the uncertainties δY_i have been underevaluated.
(c) If $\chi^2 \ll \mathcal{N} - 2$, probably the uncertainties δY_i have been overevaluated.

6. *Evaluation of g from the table.* For each pair of values $(\ell_i \pm \delta \ell, T_i \pm \delta T)$, calculate g_i through (E.45) and evaluate the uncertainty δg_i as

$$(\delta g)^2 \simeq \left(\frac{\partial g}{\partial \ell}\right)^2 (\delta \ell)^2 + \left(\frac{\partial g}{\partial T}\right)^2 (\delta T)^2$$

$$\simeq \left(\frac{2\pi}{T}\right)^4 (\delta \ell)^2 + \left(\frac{8\pi^2 \ell}{T^3}\right)^2 (\delta T)^2 \,. \tag{E.63}$$

Which one of the two uncertainties, of length and of period, has the strongest influence on the uncertainty of g?

Check if there is a possible systematic dependence of the measured value g on the pendulum length, by plotting the values g_i against the corresponding values ℓ_i.

To obtain a unique final value $g \pm \delta g$, consider and critically compare the following two procedures:

(a) Calculate the weighted average:

$$g = \frac{\sum_i g_i w_i}{\sum_i w_i}, \quad \delta g = \frac{1}{\sqrt{\sum_i w_i}}, \quad \text{where } w_i = \frac{1}{(\delta g_i)^2} \,. \tag{E.64}$$

(b) Calculate the mean $\langle g \rangle$ of the distribution of the g_i values, and estimate the standard deviation of the sample means $\sigma[\langle g \rangle]$, so that $g \pm \delta g = \langle g \rangle \pm \sigma[\langle g \rangle]$.

7. *Evaluation of g from the graph.* The relation between period and length has been expressed in (E.46) as

$$T = a\,\ell^b \,, \tag{E.65}$$

and the values $a \pm \delta a$ and $b \pm \delta b$ have been obtained from the logarithmic plot by linear regression.

In the approximation of small oscillations, the period is

E.5 Simple Pendulum: Dependence of Period on Length

$$T = 2\pi \sqrt{\ell/g}, \tag{E.66}$$

so that $a = 2\pi/\sqrt{g}$. The value of g can thus be obtained from a:

$$g = (2\pi/a)^2, \quad \delta g = \left|\frac{dg}{da}\right| \delta a = \frac{8\pi^2}{a^3} \delta a. \tag{E.67}$$

Compare the values $g \pm \delta g$ obtained by the two different procedures, from the table and from the graph.

8. *Comparison with the tabulated value.* Compare the value of g obtained in this experiment with the value available in the literature for the site of the experiment, and evaluate if the discrepancy is consistent with the experimental uncertainty (Sect. 9.3).

If the experimental value is inconsistent with the tabulated one, try to evaluate the systematic error and investigate its origin. In particular, try to understand if the systematic error of g can depend on systematic errors in the measurements of length and period, and which one of the two measurements gives the largest contribution.

Discussion

9. *Logarithms.* Two types of logarithms are mainly utilized:

(a) Decimal logarithms, with base 10:

$$x = \log y \quad \text{if } y = 10^x; \tag{E.68}$$

(b) Natural logarithms, with base the irrational number $e = 2.7182\ldots$:

$$x = \ln y \quad \text{if } y = e^x. \tag{E.69}$$

Natural and decimal logarithms are connected by:

$$\log y = \log(e) \ln y = 0.434 \ln y, \quad \ln y = \ln(10) \log y = 2.3 \log y. \tag{E.70}$$

The derivatives of natural and decimal logarithms are

$$\frac{d(\ln y)}{dy} = \frac{1}{y}, \quad \frac{d(\log y)}{dy} = \frac{0.434}{y}. \tag{E.71}$$

The function $y = ax^b$ can be linearized indifferently using decimal logarithms

$$\log y = \log a + b \log x, \quad \text{say } Y = A + bX, \tag{E.72}$$

or natural logarithms

$$\ln y = \ln a + b \ln x, \quad \text{say } Y' = A' + bX'. \tag{E.73}$$

The parameters A and A' are, however, different in the two cases:

$$A = \log a, \quad A' = \ln a = 2.3 A. \tag{E.74}$$

E.6 Simple Pendulum: Influence of Mass and Amplitude

Aims of the Experiment

- Study of the dependence of period on oscillation amplitude
- Study of the dependence of period on time
- Study of the dependence of period on mass

Materials and Instruments

- Several bodies of different masses
- A tape-line (resolution 1 mm)
- A caliper (resolution 0.05 mm)
- A stopwatch (resolution 0.01 s)
- A mechanical balance (resolution 0.1 g)

Introduction

By exactly solving the equation of motion of the simple pendulum, one finds that the period T depends on length ℓ, acceleration of gravity g, and amplitude of oscillation θ_0, according to

$$T = 2\pi \sqrt{\frac{\ell}{g}} \left[1 + \frac{1}{4} \sin^2\left(\frac{\theta_0}{2}\right) + \frac{9}{64} \sin^4\left(\frac{\theta_0}{2}\right) + \cdots \right]. \tag{E.75}$$

The period does not depend on mass. Within the approximation of small oscillations, say when θ_0 is small, one can truncate (E.75) at the first term:

$$T = 2\pi \sqrt{\ell/g}. \tag{E.76}$$

The aims of this experiment are to measure the dependence of the period on the amplitude of oscillation and to verify its independence of mass.

Experiment

1. Measurement of the amplitude. The value of the amplitude θ_0 can be obtained, by simple trigonometric relations, from several length measurements. Let us consider two different procedures (Fig. E.3); in both cases, one measures the length ℓ, then:

(a) One measures the vertical distances, from a given horizontal plane, of the suspension point and of the center of mass, h and d, respectively, so that the amplitude is

$$\theta_0 = \arccos x, \quad \text{where } x = (h - d)/\ell; \tag{E.77}$$

(b) One measures the horizontal distance s of the center of mass from the vertical line including the suspension point, so that the amplitude is

$$\theta_0 = \arcsin y, \quad \text{where } y = s/\ell. \tag{E.78}$$

E.6 Simple Pendulum: Influence of Mass and Amplitude

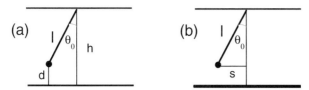

Fig. E.3. Two different procedures for measuring the amplitude θ_0.

2. *Uncertainty of the amplitude.* Evaluate the uncertainties of the length values, $\delta\ell$, δd, δh, and δs. Propagate the uncertainties to θ_0 for the two procedures:

(a) $(\delta\theta_0)^2 = \left(\dfrac{\partial\theta_0}{\partial h}\right)^2 (\delta h)^2 + \left(\dfrac{\partial\theta_0}{\partial d}\right)^2 (\delta d)^2 + \left(\dfrac{\partial\theta_0}{\partial \ell}\right)^2 (\delta\ell)^2 ,$

(b) $(\delta\theta_0)^2 = \left(\dfrac{\partial\theta_0}{\partial s}\right)^2 (\delta s)^2 + \left(\dfrac{\partial\theta_0}{\partial \ell}\right)^2 (\delta\ell)^2 .$ \hfill (E.79)

Taking into account that

$$\dfrac{d}{dx}(\arccos x) = -\dfrac{1}{\sqrt{1-x^2}}, \quad \dfrac{d}{dy}(\arcsin y) = \dfrac{1}{\sqrt{1-y^2}}, \tag{E.80}$$

one can verify that

(a) $(\delta\theta_0)^2 = \dfrac{1}{(1-x^2)\,\ell^2}\left[(\delta h)^2 + (\delta d)^2 + x^2\,(\delta\ell)^2\right],$

(b) $(\delta\theta_0)^2 = \dfrac{1}{(1-y^2)\,\ell^2}\left[(\delta s)^2 + y^2\,(\delta\ell)^2\right],$ \hfill (E.81)

where x and y are defined in (E.77) and (E.78), respectively. Analyze the variation of $\delta\theta_0$ as a function of θ_0 for the two cases (a) and (b), and check which one of the two methods is the best suited to minimize $\delta\theta_0$ for different θ_0 ranges.

3. *Damping of oscillations.* The amplitude of oscillation decreases with time, due to friction effects (mainly air friction). For a given value θ_0 of amplitude, the measurements of the period should be performed within a time interval sufficiently short that the amplitude does not significantly vary with respect to its nominal value.

Measure the number of oscillations corresponding to a given reduction of the amplitude, for both the maximum and minimum amplitudes to be considered in the experiment; evaluate the variation of amplitude for one oscillation in both cases. Estimate the corresponding contribution to the uncertainty $\delta\theta_0$, compare it with the uncertainty (E.81), and, if necessary, take it into account.

4. *Measurements of period.* Measure the period T for different values of amplitude (typically about ten values, uniformly distributed between 5 and 70 degrees). Repeat the measurements, to evaluate the effect of random fluctuations.

Plot the results on a graph with amplitude θ_0 and period T on the horizontal and vertical axis, respectively. Pay attention to specify the measurement units for amplitudes (degrees or radians). Visually check if a regular behavior is evident in the dependence of the period on amplitude. To amplify this behavior, it is convenient to determine the period T_0 corresponding to small amplitudes, and plot the difference $T - T_0$ against amplitude.

5. *Comparison with theory.* Plot T against θ_0 according to (E.75), and evaluate the relative weights of the different terms of the series expansion. Compare the experimental points with the theoretical expectation. Verify, using the χ^2 test, if the uncertainties have been correctly evaluated. Verify if there are contributions of systematic origin to the discrepancy between theory and experiment.

6. *Dependence of period on mass.* Measure the masses of the different bodies available. Measure the period for the different masses, taking care to maintain unaltered the length of the pendulum and the amplitude of oscillation. Verify the consistency of the experimental results with the expectation that the period is independent of mass, using the concepts introduced in Sect. 9.3.

Discussion

7. *Comparison of spring and pendulum.* For a body attached to a spring, the kinematical variable is the elongation x with respect to the equilibrium position, and the equation of motion is

$$m \frac{d^2 x}{dt^2} = -kx . \tag{E.82}$$

For a pendulum, the kinematic variable is the curvilinear coordinate s along the circular trajectory, or equivalently the angle $\theta = s/\ell$, and the equation of motion is

$$m\ell \frac{d^2 \theta}{dt^2} = -mg \sin\theta . \tag{E.83}$$

The gravitational mass on the right-hand side and the inertial mass on the left-hand side simplify, so that (E.83) reduces to

$$\ell \frac{d^2 \theta}{dt^2} = -g \sin\theta . \tag{E.84}$$

Only for a small oscillation, when $\sin\theta \simeq \theta$, can (E.84) be reduced to a form equivalent to the equation of a spring:

$$\ell \frac{d^2 \theta}{dt^2} = -g\theta . \tag{E.85}$$

8. *Simple pendulum and physical pendulum.* A simple pendulum is a material point, suspended by means of a massless and inextensible string. As such, it is an ideal device. A physical (or real) pendulum is a rigid body, of finite size, suspended from a fixed point. Its equation of motion is

$$I \frac{d^2\theta}{dt^2} = - mg\ell \sin\theta , \quad \text{(E.86)}$$

where I is the momentum of inertia with respect to the rotation axis, m is the total mass, and ℓ is the distance of the center of mass from the suspension point. For small oscillations, (E.86) becomes

$$I \frac{d^2\theta}{dt^2} = - mg\ell\, \theta , \quad \text{(E.87)}$$

and the period is

$$T = 2\pi \sqrt{I/mg\ell} . \quad \text{(E.88)}$$

For a simple pendulum, $I = m\ell^2$, and (E.88) reduces to (E.76). Treating a real pendulum as a simple pendulum represents a systematic error, whose extent can be estimated by comparing the two expressions of the period, (E.88) and (E.76), respectively.

E.7 Time Response of a Thermometer

Aims of the Experiment

- Measurement of the time constant of a thermometer
- Comparison between an analog and a digital thermometer (Sect. 3.2)
- Introduction to the dynamical behavior of instruments (Sect. 3.5)

Materials and Instruments

- A mercury-in-glass thermometer (measurement range from -10 to $+100°C$, resolution $0.1°C$)
- A digital thermometer (resolution $0.1°C$)
- A device for heating the thermometers
- A stopwatch (resolution $0.01\,\text{s}$)
- A water can

Introduction

Temperature θ is a nonadditive quantity (Sect. 1.2), and its measurement is based on the measurement of another quantity $\mathcal{G}$, whose dependence on temperature is assumed to be known. In this experiment, two different thermometers are compared (Fig. 3.5 in Sect. 3.3):

(a) A mercury-in-glass thermometer: the quantity $\mathcal{G}$ is the volume of a given amount of mercury; when θ increases, the volume increases and mercury is forced to flow up a thin pipe; the height h of mercury in the pipe is directly read as temperature on a calibrated scale.

(b) A digital thermometer: $\mathcal{G}$ is an electrical quantity, for example, the resistance R of a semiconductor device (thermistor): the variations of θ induce variations of $\mathcal{G}$, that are converted to digital form and can be directly read as temperature on a calibrated digital display.

The measurement of temperature requires that the probe of the thermometer attain thermal equilibrium with its environment. This process takes a finite amount of time. The aim of this experiment is to study and compare the response of the two thermometers to a sudden variation of temperature, in two different systems, air and water.

Experiment

1. Mercury-in-glass thermometer, cooling in air. Heat the thermometer up to about $\theta_0 \simeq 80°C$, then quickly extract the bulb from the heat source and let it cool down in air (temperature θ_a). Measure the temperature as a function of time during the cooling process, and evaluate the uncertainties of time and temperature. Repeat the heating and cooling cycle several times.

2. Analysis of the time–temperature graph. Plot the experimental results on a linear graph, with time and temperature on the horizontal and vertical axis, respectively. Temperature decreases with an exponential-like behavior, and asymptotically tends to the room temperature θ_a for $t \to \infty$. Evaluate θ_a and its uncertainty $\delta\theta_a$ from the graph.

3. Exponential behavior. To check the exponential behavior, it is convenient to plot the natural logarithm of $(\theta - \theta_a)/(\theta_0 - \theta_a)$ against time t (θ_0 is the temperature at $t = 0$). In the semilogarithmic graph, the experimental points should be consistent with a linear behavior, with negative slope:

$$\ln \frac{\theta - \theta_a}{\theta_0 - \theta_a} = -Bt, \qquad (E.89)$$

where B is a positive constant. The linear behavior (E.89) corresponds to an exponential dependence of temperature on time:

$$(\theta - \theta_a) = (\theta_0 - \theta_a)\,e^{-t/\tau} \quad (e = 2.71828\ldots), \qquad (E.90)$$

where $\tau = 1/B$ is the time constant of the thermometer, and is measured in seconds. The exponential behavior is consistent with a mathematical model described below (see also Example 3.26 in Sect. 3.5).

4. *Evaluation of the time constant.* Let us introduce the new variable

$$y = \ln \frac{\theta - \theta_a}{\theta_0 - \theta_a}, \tag{E.91}$$

so that (E.89) simplifies to

$$y = -Bt. \tag{E.92}$$

The uncertainty of θ, θ_a, and θ_0 is propagated to y,

$$\delta y \simeq \sqrt{\left(\frac{\partial y}{\partial \theta}\delta\theta\right)^2 + \left(\frac{\partial y}{\partial \theta_a}\delta\theta_a\right)^2 + \left(\frac{\partial y}{\partial \theta_0}\delta\theta_0\right)^2}, \tag{E.93}$$

leading to

$$(\delta y)^2 \simeq \frac{(\delta\theta)^2}{(\theta - \theta_a)^2} + \frac{(\delta\theta_0)^2}{(\theta_0 - \theta_a)^2} + \left[\frac{\theta_0}{(\theta_a - \theta)(\theta_0 - \theta_a)}\right]^2 (\delta\theta_a)^2. \tag{E.94}$$

The value of B can now be obtained by linear regression. Verify preliminarily if the uncertainties δt_i are negligible with respect to δy_i, or if it is necessary to evaluate their additive contribution to δy_i, according to the procedure of Sect. 10.4. By best fitting the function $y = -Bt$ to the experimental points (t_i, y_i), one obtains:

$$B = -\frac{\sum_i w_i y_i t_i}{\sum_i w_i t_i^2}, \quad \delta B = \sqrt{\frac{1}{\sum_i w_i t_i^2}}, \quad \text{where } w_i = \frac{1}{(\delta y_i)^2}. \tag{E.95}$$

Check the soundness of the hypothesis $y = -Bt$ by the chi square test. If the hypothesis is reasonable, evaluate the time constant

$$\tau = 1/B, \quad \delta\tau = \delta B/B^2. \tag{E.96}$$

5. *Mercury-in-glass thermometer, cooling in water.* Repeat the experiment by cooling the thermometer in water instead of in air. Again, analyze the dependence of the measured temperature θ on time, and check if experimental data are still consistent with an exponential behavior. If so, evaluate the time constant and its uncertainty.

6. *Experiments with the digital thermometer.* Repeat both experiments (cooling in air and in water) with the digital thermometer. The digital instrument samples the temperature at fixed time intervals, whose length Δt represents the minimum unit for time measurements. In evaluating the uncertainty of temperature values, one should take into account, in addition to the resolution of the instrument, also the uncertainty quoted by the manufacturer in the operation manual.

Analyze the experimental data as was done for the mercury-in-glass thermometer. Again check if experimental data for cooling in air and in water are consistent with an exponential behavior, and, if so, evaluate the corresponding time constants and their uncertainties.

7. *Comparison of thermometers.* Compare the behavior of the two thermometers, and in particular their time constants. Represent the dependence of temperature on time for both thermometers and for cooling both in air and in water on one graph, plotting the normalized time t/τ on the horizontal axis, and the normalized temperature $(\theta - \theta_a)/(\theta_0 - \theta_a)$ on the vertical axis. If the exponential model is valid, the experimental data for the different cases are consistent with the same curve.

Discussion

8. *Empirical temperatures and absolute temperature.* There are various different empirical scales of temperature, depending on the different choices of thermometric substance, thermometric property, and calibration. A well-known empirical scale is the Celsius scale, where the values $0°C$ and $100°C$ correspond to the fusion and ebullition points of water at atmospheric pressure, respectively. When measured with reference to an empirical scale, temperature is frequently indicated by θ.

The absolute thermodynamic temperature T is defined with reference to the efficiency of the ideal Carnot cycle. Its unit is the kelvin (symbol K). The relation between the values measured in Kelvin and in degrees is

$$T(K) = \theta(°C) + 273.15 \,. \tag{E.97}$$

Both scales (Celsius and Kelvin) are centigrade.

9. *A mathematical model of the thermometer.* The exponential time response of a thermometer can be described, to a first approximation, by a simple mathematical model (Sect. 3.5). Let us focus our attention on the mercury-in-glass thermometer (Example 3.26), and let be

θ the value of temperature read on the thermometer
θ_{in} the value of the ambient temperature (air or water)

The difference between θ and θ_{in} causes a flow of heat Q from the surrounding medium to the bulb of the thermometer, according to

$$\frac{dQ}{dt} = -k \frac{S}{d} (\theta - \theta_{in}) \,, \tag{E.98}$$

where t is time, k is a parameter depending on thermal conductivity of both the bulb glass and the medium, S is the bulb surface, and d is the glass thickness. The heat quantity Q is considered positive when entering the thermometer bulb, so that the heat flow is negative when $\theta > \theta_{in}$, and viceversa.

The heat transfer causes a variation of the bulb temperature θ according to

$$dQ = C \, d\theta \,, \tag{E.99}$$

where C is the heat capacity of mercury. By equating dQ in (E.98) and (E.99), one obtains a differential equation connecting the input value θ_{in} to the output value θ:

$$C\frac{d\theta}{dt} + k\frac{S}{d}\theta = k\frac{S}{d}\theta_{in} \ . \tag{E.100}$$

Because (E.100) is a first-order equation (with constant coefficients), the model of the instrument is said to be first-order.

The presence of $C\,(d\theta/dt)$ in (E.100) means that the variations of the input value θ_{in} cannot be instantaneously reproduced by the output values θ. A variation of θ_{in} will at first cause the onset of the term $C\,(d\theta/dt)$; the smaller is the thermal capacity C, the larger is the derivative $(d\theta/dt)$, and as a consequence the faster is the fitting of the output value θ to the variations of θ_{in}.

The solution $\theta(t)$ of (E.100) depends on the input function $\theta_{in}(t)$. For a step input (such as in this experiment)

$$\theta_{in}(t) = \begin{cases} \theta_0 \text{ for } t < 0 \ , \\ \theta_a \text{ for } t \geq 0 \ , \end{cases} \tag{E.101}$$

the solution $\theta(t)$ of (E.100) is

$$\theta(t) = (\theta_0 - \theta_a)\,e^{-t/\tau} + \theta_a \ , \tag{E.102}$$

where

$$\tau = \frac{Cd}{kS} \ . \tag{E.103}$$

Suggested Reading

- W. Feller: *An Introduction to Probability Theory and its Applications*, 3rd edn (John Wiley & Sons, 1968)
- H. Ventsel: *Théorie des Probabilités* (Editions MIR, Moscow, 1977)
- B. V. Gnedenko: *Theory of Probability*, 6th edn (CRC, 1998)
- A. Papoulis and S. U. Pillai: *Probability, Random Variables and Stochastic Processes*, 4th edn. (McGraw-Hill Higher Education, 2002)
- M. R. Spiegel, J. J. Schiller, and R. Alu Srinivasan: *Schaum's Outline of Probability and Statistics*, 2nd edn. (McGraw-Hill, 2000)

Data Analysis

- J. L. Stanford and S. B. Vardeman: Statistical Methods for Physical Science. In: *Methods of Experimental Physics*, vol. 28 (Academic Press 1994)
- J. R. Taylor: *An Introduction to Error Analysis: The Study of Uncertainties in Physical Measurements*, 2nd edn. (University Science Books, 1997)
- G. Cowan: *Statistical Data Analysis* (Clarendon Press, Oxford, 1998)
- P. Bevington and D. K. Robinson: *Data Reduction and Error Analysis for the Physical Sciences*, 3rd edn. (McGraw-Hill Higher Education, 2003)
- S. Rabinovich: *Measurement Errors and Uncertainties: Theory and Practice*, 3rd edn. (Springer, 2005)
- L. Kirkup and R. B. Frenkel: *An Introduction to Uncertainty in Measurements*, (Cambridge University Press, 2006)

Systems of Units, Fundamental Constants

- R. A. Nelson: *Foundations of the international system of units (SI)*, The Physics Teacher, December (1981), 596
- Bureau International des Poids et Mesures: *The International System of Units (SI)*, 7th edn. (1998)
- E. A. Cohen, K. M. Crowe, J.W.M. Dumond: *The Fundamental Constants of Physics*, (Interscience, 1957)
- P. J. Mohr and B. N. Taylor: *CODATA Recommended Values of the Fundamental Physical Constants: 1998*, Rev. Mod. Phys. **72**, 351 (2000)

Web Sites

- BIPM, Bureau International des Poids et Mesures:
  http://www.bipm.fr
- The NIST reference on constants, units, and uncertainty:
  http://physics.nist.gov/cuu/index.html

Index

Ångström, 19

Accuracy, 74
   of instruments, 35
Addition of events, 87
Ampere, 17, 219
Amplifiers, 29
Analog instruments, 30
Astronomic unit, 19
Atomic mass unit, 19
Average value of a distribution, 108
Axiomatic definition of probability, 91

Base quantities, 13
Bernoulli distribution, 100
Binomial distribution, 99, 100
   for histogram columns, 103
   mean, 113
   moments, 245
   normalization, 100
   variance, 113
Bivariate distributions, 133
Breit–Wigner distribution, 130
British systems of units, 20, 224

Candela, 17, 219
Cauchy–Lorentz distribution, 130, 252
   FWHM, 131
   mode, 131
Central limit theorem, 126
Central moments, 111, 244
Certain event, 82
cgs systems of units, 18, 226
Characteristic function, 244
Chauvenet criterion, 176
Chi square, 195
Chi square distribution, 198, 235
Classical probability, 83

Classificatory method, 4
Combinations, 95
Combinatorial calculus, 94
   combinations, 95
   dispositions, 94
   permutations, 95
Combined uncertainty, 155
Comparative method, 4
Complex events, 89
Composition of uncertainties, 72
Conditional probability, 91
Confidence interval, 171
Confidence level, 171
Consistent estimator, 148
Constraints, 197
Continuous random variable, 104
   distribution, 105
Contrary events, 88
Cosmic rays, 118
Counters, 43
Counting statistics, 118
Covariance, 136
Coverage factor, 170, 174
Coverage probability, 171
Cumulative distribution function, 105

Degrees of freedom, 147, 173, 192, 197
Derived quantities, 13
Differential equations, 239
Digital instruments, 30
Dimensional analysis, 20
   deduction of physical laws, 23
   physical similitude, 24
   test of equations, 22
Dimensional homogeneity, 22
Dimensionless quantities, 21
Dimensions of physical quantities, 21
Dirac system of units, 20

Direct measurement, 6
Discrepancy, 63
Discrete random variable, 104
  distribution, 104
Dispositions, 94
Distribution
  Bernoulli, 100
  binomial, 99, 100
  Breit–Wigner, 130
  Cauchy–Lorentz, 130
  chi square, 198, 235
  Gauss, 121
  linear correlation coefficient, 237
  Lorentz, 130
  normal, 121, 229
  Poisson, 115
  Student, 173, 233
Distribution of sample means, 58
Distributions
  average value, 108
  bivariate, 133
  central moments, 111
  expected value, 108
  FWHM, 110
  initial moments, 111
  mean, 108
  median, 109
  mode, 109
  moments, 111
  multivariate, 132
  numerical characteristics, 108
  of continuous random variable, 105
  of discrete random variable, 104
  position parameters, 108
  skewness, 112
  standard deviation, 109
  variance, 109
Distributions of sample means, 144, 145
Distributions of sample variances, 144, 146
Double probability density, 134
Dynamical behavior of instruments, 37

Effective estimator, 148
Electronvolt, 19
Error Function, 125
Estimation of parameters, 57, 147
Estimators, 148
Eulero–Poisson integral, 122, 248

Event, 81
  certain, 82
  impossible, 82
  probability of an event, 82
Events
  complex, 89
  contrary, 88
  independent, 93
  mutually exclusive, 88
  product of, 87
  sum of, 87
Expanded uncertainty, 174
Expected value, 108
Experimental method, 12
Extended uncertainty, 71

Frequency, 83
Full-width at half maximum, 110
Functional elements of instruments, 27
Fundamental constants, 227
FWHM, 110

Gauss distribution, 121, 229
Gaussian distribution, 55
Geiger counter, 43
Graphs, 212
  linear scales, 213
  logarithmic scales, 214
  semilog scales, 214
Greek alphabet, 227

Half-width at half maximum, 111
Hartree system of units, 20
Histograms, 50, 216
  area normalized, 51, 217
  height normalized, 50, 217
  limiting histogram, 54
  mean, 52
  standard deviation, 52
  statistical parameters, 51
  variance, 52

Impossible event, 82
Independent events, 93
Indirect measurement, 8, 155
Initial moments, 111, 244
Instruments, 27
International system of units, 15, 219
  base units, 15, 219

derived units, 18, 220
   prefixes, 222
Interpolation, 211

Kelvin, 17, 219
Kilogram, 16, 219

Least squares, 186
Likelihood function, 149
Limiting distribution, 54
Limiting histogram, 54
Linear correlation coefficient, 179, 237
Linear interpolation, 211
Linear regression, 182
Linearity of instruments, 32
Lorentz distribution, 130

Marginal probability, 134
Mathematical models of instruments, 38
Maximum likelihood, 149, 187
Mean
   parent mean, 143
   sample mean, 52, 143
Mean of a distribution, 108
Measure, 6
Measurement, 6
   direct, 6
   indirect, 8, 155
Measurement range, 31
Measurement resolution, 46
Measurement standards, 14
   artificial, 15
   natural, 15
   properties, 14
Measuring chain, 28
   closed chain instruments, 30
   open chain instruments, 29
Measuring instruments, 27
   absolute, 29
   accuracy, 35
   active, 31
   analog, 30
   differential, 29
   digital, 30
   displaying, 31
   dynamical behavior, 37
   linearity, 32
   mathematical models, 38

measurement range, 31
operating conditions, 34
passive, 31
recording, 31
resolution, 33
sensitivity, 32
transparency, 34
Median, 109
Meter, 16, 219
Metric system, 14
Minimization
   linear functions, 188
   non linear functions, 189
Mode, 109
Mole, 16, 219
Moment generating function, 243
Moments, 111
   binomial distribution, 245
   central, 111
   initial, 111
   normal distribution, 248
   Poisson distribution, 247
   uniform distribution, 246
Morphological method, 3
Multiplication of events, 87
Multivariate distributions, 132
   continuous random variables, 134
   covariance, 136
   discrete random variables, 133
   independent random variables, 135
   marginal probability, 134
Mutually exclusive events, 88

Natural systems of units, 20
Newton binomial, 96
Nonstatistical uncertainty, 69
Normal distribution, 55, 121, 229
   central moments, 123
   FWHM, 123
   kurtosis, 123
   mean, 123
   moments, 248
   normalization, 122
   standard deviation, 123
   standard variable, 124
   variance, 123

Operating conditions of instruments, 34

Operative definition of physical
  quantities, 11

Parent mean, 143
Parent population, 139
Parent variance, 143
Permutations, 95
Physical quantities, 5
  additive, 5
  base quantities, 13
  derived quantities, 13
  nonadditive, 6
  operative definition, 11
  time dependence, 8
Physical quantity, definition, 4
Poisson distribution, 115
  counting statistics, 118
  for histogram columns, 121
  kurtosis, 116
  mean, 115
  moments, 247
  skewness, 116
  variance, 115
Poisson processes, 118
Polynomial regression, 189
Position parameters of distributions, 108
Practical systems of units, 19
Probability, 82
  axiomatic definition, 91
  classical, 83
  conditional, 91
  of the product of events, 91
  of the sum of events, 89
  statistical, 83
  subjective, 84
Probability density, 106
Probability interval, 170
Product of events, 87
  probability of, 91
Propagation of uncertainty
  difference, 158
  direct proportionality, 159
  general expression, 161
  linear functions, 156
  maximum uncertainty, 167
  nonindependent quantities, 163
  nonlinear functions, 159
  product, 162
  quotient, 163
  raising to power, 161
  sum, 158

Quantitative method, 5

Radian, 18
Random errors, 36, 48
Random events
  and physical quantities, 10
  counting of, 10, 43
Random fluctuations, 36, 48
Random phenomena, 79
Random variables, 104
  continuous, 104
  discrete, 104
Reduced chi square, 201
Regression
  linear, 182
  polynomial, 189
Rejection of data, 175
Relative uncertainty, 73
Repeatability, 73
Repeated trials, 99
Reproducibility, 73
Resolution of instruments, 33
Resolution of measurement, 46
Rounding, 208

Sample covariance, 179
Sample mean, 52, 143
Sample population, 140
Sample space, 81
Sample standard deviation, 52
Sample variance, 52, 143
Schwartz inequality, 166
Second, 219
Sensitivity of instruments, 32
SI, 15
Significant digits, 207
Skewness, 112
Stability, 36
Standard deviation, 109
  sample standard deviation, 52
Standard normal density, 124
Standard normal variable, 124
Statistical frequency, 83
Statistical independence, 156
Statistical inference, 147

Statistical methods, 5
Statistical probability, 83
Statistical uncertainty, 69
Steradian, 18
Student distribution, 173, 233
Subjective probability, 84
Sum of events, 87
  probability of, 89
Systematic errors, 36, 61
Systems of measurement units, 14
  british, 20
  cgs, 18
  Dirac, 20
  Hartree, 20
  natural, 20
  practical, 19
  SI, 15

Tables, 210
Transducers, 29
Transparency of instruments, 34

Unbiased estimator, 148
Uncertainty
  causes of, 45
  combined, 155
  due to random fluctuations, 59
  due to resolution, 47
  due to systematic errors, 67
  expanded, 174
  extended, 71
  in direct measurements, 7, 45
  non statistical, 69
  relative, 73
  statistical, 69
  Type A, 69
  Type B, 69
  unified expression, 69
Uniform distribution
  mean, 114
  moments, 246
  variance, 114
Unit standard, 6

Variance, 109
  parent variance, 143
  sample variance, 52, 143

Weighted average, 66, 150